Human Evolution and Male Aggression

Human Evolution and Male Aggression

Debunking the Myth of Man and Ape

Anne Innis Dagg and Lee E. Harding

CAMBRIA PRESS

Amherst, New York

Requests for permission should be directed to
permissions@cambriapress.com, or mailed to:
Cambria Press
20 Northpointe Parkway, Suite 188
Amherst, NY 14228

Library of Congress Cataloging-in-Publication Data

Dagg, Anne Innis.
Human evolution and male aggression : debunking the
myth of man and ape / Anne Innis Dagg and Lee Harding.
p. cm.
Includes bibliographical references and index
ISBN 978-1-60497-821-6 (alk. paper)
1. Human evolution. 2. Social evolution. 3. Human behavior.
4. Inter-male aggression. I. Harding, L. (Lee) II. Title.

GN281.D35 2012
599.93'8—dc23

2012026268

To the late Ian Ralph Dagg
and to Andrew Harding and Jill Harding

Table of Contents

List of Figures

PREFACE

While researching literature for a paper on silvered leaf monkeys, a Malaysian species, Lee Harding discovered that infanticide by males, long reported as a common behavior of this species, had never actually been observed. In the 1970s, the original author of the report had assumed that it might have occurred because of a new theory then sweeping the nascent field of sociobiology: that when taking over a troop of monkeys or a pride of lions, males often killed the infants, ostensibly to bring the female into estrus so that he could mate with her. The theory had been put forward in 1974 based on observations of hanuman langurs, a South Asian species of leaf monkey, as well as on solid biology: Primate females who are nursing do not ovulate: they come into estrus, and hence become willing to mate, only after their infants are weaned or killed. Later research confirmed that infanticide has adaptive advantages in some hanuman langurs (Borries, Launhardt, Epplen, Epplen, & Winkler, 1999). However, Harding realized that the biology of silvered leaf monkeys precluded any such benefit to the males, for reasons explained in chapter 10.

Knowing that Anne Innis Dagg had been involved in the scientific controversy about male infanticide—whether and how often it occurred, whether it could have an evolutionary basis—that was debated in journals for several years (e.g., Dagg, 2000), Harding asked her about the practice in silvered leaf monkeys. Although she had not studied this species, Dagg told him that the hypothesis had become so pervasive in sociobiology that it was often assumed without evidence. We wondered what other errors of observation or interpretation might have biased the scientific and popular understanding of male aggression in primates. We began to turn up quite a list.

We found, for example, that the male infanticide hypothesis had been misapplied to the 29 or so species of Southeast Asian leaf monkeys and 16 species of gibbons (and by extension to most other primates). Their biology precludes any possible benefit of infanticide, which, in fact, has never been observed in most of these species.

While the infanticide fallacy was sweeping sociobiology, another fascinating finding was creeping into the professional and public consciousness in the 1970s: New genetic analysis techniques had shown that human DNA is 98.9% identical to that of chimpanzees. Anthropologists rushed to equate human behavior with that of this species, and some taxonomists even proposed moving chimpanzees into humans' own genus, *Homo*.

Several long-running chimpanzee studies by Jane Goodall and others were particularly damning because they documented many cases of infanticide and other homicides in humans' closest living relative which had nothing to do with the hormonal cycles of females. These ideas—that males are inherently aggressive and infanticidal, and that they are rewarded for killing infants other than their own by (a) having more sex and (b) passing on more genes—not only took sociobiology by storm, but captured the public imagination. Here, it seemed, was an evolutionary explanation for certain otherwise disagreeable behaviors in the human species: Men are inherently aggressive, even murderous, and use every means, no matter how heinous, to spread their genes. The idea explained a lot: wars, adultery, violence against women. It was, after all, just boys being boys.

The kicker came in 2010 when primatologists Robert Sussman and Joshua Marshack (2010) reviewed all the data of three major, long-running chimpanzee studies and found infanticide to be "extremely rare"—there were 17 instances, to be exact, in 215 total years of observation. They concluded that "the data for infanticide...are so sparse...[that] both the evidence and the interpretations are suspect" (pp. 23–24).

This got us to wondering about human beings. New advances in human evolution have shown that contrary to the 1970s thinking of human ancestors as violent nomadic groups whose males fought each other over females, it is now known that they lived in pair-bonded or monogamous, relatively peaceable societies for at least the last 4 million years. And, they were nothing like chimpanzees. For example, contrary to the popular view, humans' ancestors never knuckle-walked. Chimpanzees and gorillas developed that mode of locomotion independently, long after they had diverged from the human lineage. Human beings' DNA is 97% identical to that of orangutans, which do not knuckle-walk and whose ancestors never did, either, and no one is comparing human behavior to that of orangutans.

Have the general ideas of male aggression and the need for females to protect against it unduly influenced thinking about the evolution of the human species? Have obsolete ideas from sociobiology influenced professionals' thinking about modern human behavior? Have they influenced public perceptions? Do males get something else from monogamous relationships besides exclusive access to sex with their mate—love, perhaps, or the satisfaction of fatherhood, or a good meal? As recompense for limiting their mating opportunities, do females get something else from their mate besides an economic advantage—love, possibly, or help with child rearing?

We keep thinking about how *Homo erectus*, modern human beings' immediate predecessor, managed to walk all the way from Africa to Java, and from there to China (Rightmire, 1993). The cooperation and sharing of child rearing and household chores needed to meet the challenges of such an adventure seemed inconsistent with a relationship based on fear of aggression. This is especially so with regard to the transport of infants who, because of the delayed development mandated by humans' huge brains, took many years to mature enough to travel independently. A woman could not carry one infant and lead several toddlers and dependent children very far, especially on journeys involving crossing large rivers and navigating seas, as *H. erectus* did. She needed help, and an infanticidal male would not have been the mate of choice. She needed more than help: She needed a long-term commitment. Why would he make such a commitment? These questions were the genesis of this book.

Acknowledgments

We are grateful to the following people who have gone out of their way to help us in our work: Susan Gow, Rita Huang, Joshua Marshack, Robert Sussman, Lisa Tomalty, and Hannah Diamond. We also thank the staff of Cambria Press for their help on this book.

Human Evolution and Male Aggression

Section I

Primate Biology and Behavior

This section introduces or reviews some matters of basic biology and behavior that are needed to understand the scope and dimensions of aggression in primates. Chapter 1 introduces our subject matter and how we plan to address its challenges. Chapter 2 provides an update on recent advances in archaeology (the study of old bones and artifacts), hormones, genes, and brain evolution that seem to us to require a revision of the thinking about the evolution of human behavior. Chapter 3 discusses biases in the story of human evolution. Chapter 4 dispels antiquated, but still commonly held, notions about humans' closest relatives, chimpanzees and bonobos. In chapter 5, we review research on behaviors that forestall aggression, which are too often overlooked in discussions of how humans got to be what they are today.

Chapter 1

Introduction

Scientists are supposed to be unbiased in their research and writings, but history shows that this is not necessarily what happens. Scientists live within their cultures and take on the prejudices that prevail in their lives. Perhaps no scientific area is more rife with bias than that of human evolution. For this book—which of course we presume to be unbiased (as is usual for such books)—our specific topic considers why scholars have assumed humans' early male ancestors were aggressive rather than largely peaceful individuals living in amicable groups whose calm was now and then disrupted by squabbles. Here are four recent and/or prevailing biases on the subject of animal behavior.

Bias Against Females

Behavioral studies used to be rife with sexism. In 1983, Dagg published a book entitled *Harems and Other Horrors: Sexual Bias in Behavioral Biology*. She wrote,

> The social behavior or sociobiology of animals until the past two decades has been studied almost entirely by men who have brought with them to their work the concept that women are inferior to men.

> When these biologists watched what animals were doing, their bias against women became a bias against females: They spent more time watching and describing male behavior, and attributed more significance to it than to female behavior. Writers used such biased information to refer back to women; because female animals were reported to be passive in mating and unaggressive in general, it was concluded that women should be this way too.

Linda Marie Fedigan (1982) and Shirley Strum (1987) have also documented this problem.

Female animals were made to seem inferior to males in three ways (Dagg, 1983):

1. Animals and their relationships were described by words and concepts that were biased and represented stereotyping against females rather than fact. For example, if a few females hung out with one male, they would often be referred to as his harem despite a lack of any reproductive activity.

2. Aggression in female animals was downplayed in accordance with the presumed nonaggression of women. This perception continued long after female rather than male lions were found to be the hunters and killers of their pride, just as female hyenas are of hyena groups.

3. Females' active role in sex was denied. This was more difficult to do after Jane Goodall (1986) described the libidinous chimpanzee Flo, who in her older years and while feeling sexy had 14 males trailing after her for sex on one occasion, and who years later copulated 50 times in one day.

Now, in 2011, such sexual bias is almost gone, thank goodness. Women are at least as numerous in behavioral animal research as are men, and both men and women describe what really goes on in the wild. Women often chose different topics to study (such as those concerning females and cooperation rather than competition), but their articles are published in the most prestigious journals (Adams & Burnett, 1991).

It is now known that female animals can be as aggressive, sexy, and dominant as males and that they sometimes outdo males in these traits. However, as we note in chapter 2, popular books and articles about aggressive males written by men continue to be legion, whereas documents by women (and a few men) disputing this mind-set are few.

Bias Against Homosexuality

Bias against homosexuality has had a less obvious history, but one that is devastating to gays and lesbians, who are frequently told that homosexual behavior is "against nature" (or aberrant, abnormal, deviant, perverse, etc.)—a phrase that is patently false. In the early 1980s when Dagg (1984) carried out the first census of animals for which homosexual behavior had been described, she found many scores of examples. Bruce Bagemihl (1999) has since greatly expanded this number of species in his impressive book *Biological Exuberance*.

Why was there such a bias? It existed partly because biologists believe in evolution—there was no obvious reason why homosexuality would evolve or occur in successive generations when it had no reproductive purpose. Same-sex behavior had long been observed in zoos and research labs, but this was believed to be a consequence of captivity. When some zoologists working in the field saw males mounting other males, they persuaded themselves that this was impossible: They must have been mistaken somehow. Others accepted that such behavior happened but chose not to report it lest they be accused of being gay themselves or of having kinky ideas. When one researcher on the activities of Japanese macaques, Linda Wolfe, wanted to have her results (which included documentation of homosexual behavior) published, referees accused her of making up facts and of doctoring photographs that illustrated them (Vines, 1999).

The bias persisted also because many zoologists, like people in general, were homophobic. They did not want to concede that the animals they studied, "their" species, would behave in such a way. Valerius Geist (1975) saw many of his mountain sheep males mounting each other, but at first he called this activity "aggressosexual" rather than homosexual behavior. He wrote, "To state that the males had evolved a homosexual society was emotionally beyond me. To conceive of those magnificent beasts as 'queers'—Oh God!" and Geist repudiated "that drivel!" (pp. 97–98). He called a spade a spade and admitted that his rams did indeed live in essentially a gay fellowship.

From these examples, it is clear that homophobia was being transferred, consciously or unconsciously, by scientists from the human to the animal world, just as sexism had been. Nowadays the possible evolutionary reasons for homosexuality in animals remain problematic, but at least it is acknowledged to happen in many or most animal species as well as in human beings. Yet even today, when most cases of homosexuality in humans are known to have a biological (genetic or hormonal—not psychological) basis, bigots continue to harass and even murder people who are gay or lesbian.

BIAS AGAINST FULL PARTICIPATION OF AFRICANS

Society would have been amazed if an African in the past had come to America to study the behavior of bison. Like Jane Goodall, he or she may not have had a university education but would still have been anxious to research the large migrations of this species, which might shed light on the vast annual movements of animals in Tanzania. News media would have wondered why Americans and Canadians did not study this themselves. Radios might have blared that members of a society should set their own agenda about what is important to study.

Scientists in Western nations and in Japan were anxious to do research on African animals when peace came after the Second World War. Beginning in the mid-1950s, foreign zoologists began to trickle into the continent in a steady stream to study exotic species. Ideally, Africans would have been doing this work themselves, but they did not have the academic background or financial backing for this. It took many years for Westerners to appreciate that Africans should be fully involved in all research and should shape many of the questions that a team planned to pursue. Some Japanese were more sensitive, and one team, led by Takayoshi Kano (1992), insisted that no primatologist could come to work in their bonobo (pygmy chimpanzees, *Pan paniscus*) research station in Wamba, Zaire, until he or she had learned the native language of the area, Lingala. That way, the scientists could readily hire and train local people to help with the research and discuss their findings with the neighboring populace. Bonobos would become not bush meat to be slaughtered, but interesting animals to observe and enjoy.

There was a huge breakthrough for Africans when kidnappers raided Jane Goodall's chimpanzee station at Gombe in 1975 and took four white students hostage. Before that time, university students newly arrived from America would head field research projects and long-term employed Africans would be assigned to help them. From then on, because it was dangerous for white people to stay at Gombe, "the Tanzanian field staff moved front and center" (Peterson, 2006, p. 562). Africans who had been merely field helpers took over responsibility for the research, which they did with distinction. The compilation of field notes continued as efficiently as before: There were "A" records for observations at the feeding area and "B" records from the following of individual chimpanzees in the forests.

Since then, Africans in all parts of the continent have taken important roles in research along with white scientists. Wildlife researchers from Latin America and Asia (with the exception of Japan) have been little involved, though. It is wonderful to note that animals themselves are now often given African names, such as Makale and Mutesi for two elephant friends, because they are African too.

BIAS TOWARD MALE AGGRESSION

This book addresses a fourth bias that is widely present in human culture, that of aggression. In writing history, which was until recently done largely by men, male conquests and wars were showcased. Families want safety and serenity in their lives, but historians preferred to report on the clash of nations. A country living in peace for 200 years often became of interest to historians only after it embarked on a war.

Aggression has also biased the study of primate behavior during human evolution. The worship of male aggression in human cultures has tainted some of the research on primates and falsified the story of how human beings evolved from their early primate ancestors to the present.

TYPES OF AGGRESSION

Most animals have inherited the potential to be aggressive, which can be defined here as engaging in assaults to harm, defeat, or defend against another animal or animals in battle. Three basic types of aggression exist in social animal species. First, and probably most important, is aggression against neighboring groups that might try to annex territory or harm some members of a group. Social animals must be ready and able to fight to maintain group integrity, including being willing to lose their lives in an attack by a predator. Second, aggression is used by carnivores and a few primate species that hunt and eat meat. These two types of aggression are positive for social groups.[1] The third type, which is animals quarreling within a group, can be negative for some group members or the group itself. It undermines collegiality, destroys coalitions, and stresses group members, jeopardizing their health.

For our purpose in this book, we limit the discussion to the third type of aggression: fighting within a group, not against other groups, predators, or prey. Most such struggles involve adult males battling each other.

New Ways of Thinking

Until recently, aggression was assumed to be a major force in the evolution of many species. If a male is strong and aggressive enough to be able to mate with most of the females in a group, his genome will obviously be important to the group's future. Competition was presumed to be pervasive. *The Mermaid's Tale: Four Billion Years of Cooperation in the Making of Living Things*, by Kenneth Weiss and Anne Buchanan (2009), however, tells a different story. In the past, evolution was thought of as something large that happens over hundreds or thousands of years. No one, even Charles Darwin, knew exactly how this was carried out in the short term. Weiss and Buchanan argued that natural selection is one way, but only one way, by which evolution can proceed. Evolution is, rather, a multidimensional process.

Evolution can occur because of environmental change. Consider a northern pond surrounded by forest. In natural succession over the years, the pond slowly filled with reeds, then gradually with more terrestrial plants, and finally with saplings. Fish and frogs that lived in the pond disappeared, to be replaced by woodland creatures such as mice and shrews. One population of animals has "won" and another has "lost," but there has been no contact between them.

Another way is through organismal selection, meaning when members of a species select a new environment (although usually not consciously). Perhaps part of a forest in North America may be clear-cut, leaving areas of scrub and brush behind. Deer will soon move into this new region. Again, this could lead in the future to an evolutionary change.

Or, members of a species may spread out over a wide area and, by chance, become separated into two factions. One group may even have swum to an island while another remained on the mainland. In time, with random mutations occurring in each group, these populations will eventually be unable to interbreed, or they may not want to do so. Again, there has been no competition.

None of these changes has been brought about by reproductive fitness. There has been no "survival of the fittest." All of them have involved both life forms and ecosystems. So, it is necessary to begin to appreciate that aggression is much less important in the evolution of a species than had been thought.

Weiss and Buchanan described a second insight pertinent to a form of cooperation that is omnipresent throughout life processes. To put an anthropomorphic spin on this, sperm and ovum cooperate to become a zygote, with specific chromosomes from each lining up together in a cooperative manner. The zygote divides into cells that are similar and get along well together (as do the proteins that are neatly paired and folded within them). Eventually, certain cells form into various tissues, again all working together to carry out a particular function. And tissues themselves operate together to make the organism function smoothly. Indeed, tissue may be even more adaptable; should someone lose his or her sight, the visual areas of the brain can cooperatively be recruited to enhance the hearing or touch of the individual so that he or she will have improved function in life. As shall be seen, animals of all stripes cooperate in mating, often in other activities, and always in play. If the initial trillions of cells and many tissues do not cooperate fully during development, perhaps because of a chance mutation, the individual will likely die. Obviously, cooperation, not aggression, is the essential mainstay of life.

CONTENTS OF THIS BOOK

Following this introduction, chapter 2 summarizes evidence about the fossil remains of human ancestors from which scientists have extrapolated how they think these individuals behaved. Popular writers and journalists have added another layer of interpretation that informs—and sometimes misinforms—public understanding of this evolving science. Many men and some women have written books about early human ancestors being aggressive hunters, which are discussed (and found wanting) in chapter 3.

In chapter 4, we consider the behavior and biology of chimpanzees and bonobos, both known from DNA samples to be humans' closest relatives, and show that bonobos (which are peaceful and sexy) rather than chimpanzees (which are antsy and more aggressive) behave more like human beings, although neither ape has the humans' pair-bonded mating system. In chapter 5, we explain that although males tend to compete for mating rights with females, various appeasement behaviors have evolved in many primate species to forestall aggression within their groups.

How might monkey and ape behavior today resemble that of human forebears? Nuclear families are the social organization of most human cultures, and in chapter 6 we show that this lifestyle has also evolved in many successful primate species. In chapters 7, 8, and 9, we give examples, respectively, of the various primate species whose males are affectionate to infants, get along well with females, and become close friends with other males. We examine in chapter 10 the overemphasis or even valorization of the hypothesis of males in many species committing infanticide so that they themselves can then mate with the females to produce their own offspring. This does happen in a few species, but it is claimed for many species in which it does not occur. In chapter 11, we show that female animals are about as likely as males to commit infanticide, although this fact is little known given that it undermines the infanticide-by-males hypothesis. In chapter 12, we summarize important points made in this book and provide a new view of aggression in primates.

Variability of Social Behavior

Many male authors have argued that because men are aggressive, humans' male ancestors must have been aggressive too. However, social behavior depends on many factors and is not set in stone (Bekoff & Pierce, 2009). As an extreme example, dogs and wolves can interbreed, yet their behaviors are completely different. Wolves live in packs with an alpha male, which breeds with the alpha female and also helps to rear his pups; other adult males and females do not breed but instead help with pup rearing and provisioning duties. In contrast, male dogs do not team up with the females they impregnate or help in any way to raise their own young. They lost this behavior during domestication.

Even the hominoid family exhibits considerable variation in mating systems. Gibbons and humans, for example, have stable pair bonds with relatively infrequent extrapair copulations and live in nuclear families, whereas orangutans are solitary, gorillas have a harem system, and bonobos and chimpanzees have contrasting types of promiscuity.

Anatomical changes can be traced over millions of years in the bones of humans' distant ancestors, but inferring their activities or societies requires subjective interpretation. It is known, however, that present-day species in the wild belonging to the same genus can have very different behaviors. Witness hamadryas baboons, *Papio hamadryas*, whose males are focused entirely on their few harem females, and olive baboons, *Papio cynocephalus*, whose males mingle freely with all the troop females. Or consider chimpanzees, *Pan troglodytes*, which are much more aggressive (especially when frustrated by investigator-induced supply problems with banana offerings) than bonobos, *Pan paniscus*, which seem to live for sex of both the homo and hetero kind.

Although baboon males are usually feisty, Robert Sapolsky (2011), who has studied wild savanna baboons for almost 30 years, reports that the aggression of a group may be a cultural rather than an inherent characteristic. The fiercest males of the Forest Troop and the Garbage Dump Troop fought continually over garbage tossed out by a Kenyan hotel that in time included tubercular meat. Most of these males died of tuberculosis. The new Forest Troop as a result comprised fewer males, and they were amiable rather than belligerent. Fighting was largely replaced by mutual grooming; males even groomed each other—"a behavior nearly as unprecedented as baboons sprouting wings" (Sapolsky 2011, p. 36). Ten years later, the genial surviving males from the tubercular era had been replaced by younger immigrant males. Despite these new males, the troop retained its unique social milieu, as it continues to do 20 years later. A multigenerational benign culture had been born in the Forest Troop of baboons.

Other examples of learned social differences involve chimpanzees. In Senegal, these apes frolic about up to their chests in a pond of water from a recent rainfall, grooming and playing in what amounts to a "pool party" (McGrew, 2010). Half a continent away, chimpanzees make elaborate detours through trees or jump from rock to rock so they will not have to step even into shallow water. What is the reason for this difference?

As another example, the principal investigators of nine long-term research programs at different sites compared the cultural behaviors of their animals and found no two populations alike (Whiten, Goodall, McGrew, Nishida, Reynolds, Sugiyama, & Tutin, 2001). In addition, from their list of 39 observed behaviors, only three can be considered aggressive.[2] Even within one group, behavior over the years is constantly changing depending on varying sex ratios, age ratios, group size, food supply, and the peccadilloes of individual animals. (Human societies have extensive cultural differences—think Amish compared to hippies—but these are far more extreme because they can be perpetuated by language and written material.)

Whether male monkeys and apes are inclined to be sociable among themselves depends on a number of factors (Hill, 1994; Hill & van Hooff, 1994). Kinship may explain friendly behavior between some male individuals, but many sociable males are not related and relatives are sometimes enemies. Friendly relations are often negatively correlated with competition for such things as food and mating rights. The greater the competition within a group, the more tension and aggression there is. This is all too obvious within provisioned (artificially fed) groups. Alternately, the more competition there is between groups, the less there will be within a group. Usually, the more equal the number of adult males and females, the less tension there is within a group.

PROBLEMS IN DECIPHERING PRIMATE BEHAVIORS

When describing the behavior of various primates, we follow the lead of researchers. They have observed what individual monkeys or apes are doing, and it is they who determine what the behavior means. For example, confusion could arise because (a) a male that cares for an infant can be said to do so either because he is related to the youngster or because he is kind, (b) a male baboon playing with an infant can be said to be just an amiable character or someone acting to prevent infanticide, or (c) a male gorilla displaying toward females in his group can be said to be showing them how appealing he is or coercing them to have sex with him in the future (Robbins, 2009). Similarly, if a male supports a relative, it can be said he is doing so because they share large parts of their DNA. If he does so with an unrelated animal, it can be said he is doing this to gain an advantage. But perhaps he is simply friendly with another male because he likes and gets along with him.

Evolutionary biologists agree that stereotypic behaviors do not evolve by accident: They are there for a reason, and the reason involves the species' ability to survive, prosper, and propagate. But interpreting the adaptive significance of behavior is fraught with uncertainty and inevitably colored by the prevailing theoretical framework of the day. There are different explanations in evolutionary as opposed to proximate time, and these are often confused or ignored.

There is no "right" behavior in evolutionary terms. Each species evolves and adapts its behaviors as necessary within the constraints of its own biology and its environment. If the environment changes, or if part of the population moves outside the species' core range into a novel habitat, new behaviors may be needed, and the individuals that adapt best will pass on their genes more than others do. This is what drives speciation and provides the tremendous diversity not only of species but of variety within a species. The human ability to develop a diversity of cultures, pass them on to new generations, and adapt them rapidly to changing circumstances must be seen as a feature of biology that contributed to humans' success.

This ability is so complex as to defy easy explanation and is one that causes friction between and within societies. It is ironic that although it is possible to describe the behavior of baboons and gibbons in considerable detail, it is difficult to describe some aspects of human behavior in terms that that all people will understand and agree upon. In addition, biologists and primatologists often ignore those scholars who do specialize in human behavior, such as social anthropologists.

Our Parameters

Most of the examples in this book are taken from animals living in the wild. Only occasionally, to make a specific point, do we refer to animals confined in research laboratories or in zoos. This is because the behavior of a species can drastically change in captivity. For example, Joan Silk (1994) noted that adult male bonnet macaques in captivity (which have had their canine teeth routinely clipped and blunted) "frequently sit together, groom, huddle, greet, and support one another" (p. 288). In the wild, however, although the males still clasp together in a bunch to rest (Sugiyama, 1971), they sometimes seriously injure each other (Silk, 1994). As shall be seen in chapter 3, hamadryas baboons, by contrast, can be far fiercer in captivity than in the wild. As another example, female dominance hierarchies have never been reported in wild silvered lutungs, that most peaceful of leaf monkeys, but they do sort themselves into dominance rankings in captivity when food is provided (Amarasinghe, Botejue, & Harding, 2009). It is imperative that behavior observed in caged individuals not be assumed to be the same as that of animals in the wild. It is also worth noting that sometimes conditions in the wild have been so altered by human beings that habitats become more like bad zoos than natural environments.

Our information about fossil material is as up to date as possible, given that new finds continue to be reported. All scientific information is referenced with sources, noted in superscripts, and listed in the bibliography.

The scientific names of the animals discussed are given in an appendix, along with a glossary for unusual but necessary terms.

To conclude this introduction, and keeping in mind the insidiousness of bias, we give our own personal statements about our experiences with aggression.

Personal Statement by Dagg

I am a middle-class woman who does not visit bars but who has been out and about for many years. I have read about angry men who beat up their wives and murder their enemies. In movies I have seen thousands of men acting in rage, smashing furniture and blowing up buildings. I have seen endless competitors in ice hockey games tripping and high-sticking each other. But in real life, I have never seen a man act in a violent way toward any other person (except for spanking a child, unfortunately). Once at a party, the host called his grown-up son a "mongrel," at which the son stood up from the dinner table in anger, waving his fists. However, the host stayed seated, rolled his eyes, and the son soon sat down again. Dinner carried on. Men are more aggressive than women, but aggressiveness in no way defines them.

Personal Statement by Harding

Unlike Dagg, I have seen men fighting each other and have done so myself, but not since university. It is my guess that every boy coming of age has opportunities to either accept a challenge and fall or stand as the victor or to decline and be thought a coward. Certainly this was a near-universal experience in my junior high school and high school, but it was rare in university and never, in my own experience, occurred in adulthood.

A personal statement about male aggression would be incomplete without mentioning that I had another opportunity to fight when, as an American, I was drafted for the Vietnam War. Among my close circle of high school friends, most were drafted or signed up (one came home in a box), one was a "conscientious objector" who did equal time in an alternative service, and I immigrated to Canada. Although these decisions had more to do with an understanding of international politics and personal ethics than with aggression per se, they do bring up two important questions: If, as I believe, men are inherently peaceful, why do some people attack other men or women? Why do people band together to go to war? This book does not answer these questions, but it does provide a foundation for understanding the diversity of male aggressive behavior in primates and its evolutionary basis—or rather, the lack thereof—in humans.

ENDNOTES

1. By the time *Homo sapiens* evolved about 200,000 years ago, they were already expert hunters. Tim Flannery in his celebrated book *Here on Earth* (2010) described how human ancestors in the course of inhabiting the world outside Africa wiped out scores of large animals that are now known only from their fossils.
2. These three documented behaviors were (a) "clubbing," or striking out with a stick, sometimes used against other chimpanzees but also used against snakes, baboons, and leopards; (b) "aimed throw," or throwing an object (often inaccurately) at another chimpanzee, a leopard, or monkeys; and (c) "drag branch," or running while dragging a large branch during an aggressive display.

Chapter 2

Humans

Harding related the following incident:

> While digging in the garden, I hear a shout down the street and look up. Four houses down at an intersection, my neighbor has yelled at a car full of teen-aged boys who are driving too fast past a school. A few moments later, the car comes back, still too fast, and my neighbor yells again. The car stops with a squeal of tires and three youths get out and approach my neighbor swearing and gesturing threateningly. Without thinking, I shoulder my shovel and hurry over. As I pass the house next door, that neighbor grabs his rake in two hands and joins me. A third man from across the street comes holding a hoe. None of us says a word, but when we stand beside the man who has yelled, the kids run back to their car and leave, driving acceptably slowly.

By any measure, four men, three of them armed, facing three teenagers is an aggressive act. But context is everything. My neighbors and I had sprung to the defense not only of our neighbor but also of the kids of the school, none of whom were ours. I have often thought of this trivial incident when studying other primates. Our action was not strictly territorial defense, as is seen in many primates, because it occurred well away from our own houses on another street. It was community defense. With nothing to gain individually from scolding these errant adolescents, we knew instinctively that public safety is good for the community, and we each shouldered our weapons to uphold it. This kind of aggression, supporting the community beyond self-interest, is part of being human, but not uniquely so: We share community defense with many other primates. As for the youths, we will return to them shortly.

BONES

This is an exciting time to be a primatologist. New discoveries of ancient bones, new methods of analysis, new theories and ways to test them, and the whole new genre of DNA analysis add daily to the understanding of human evolution. The downside is that anyone whose last biology course was more than 10 years ago is likely to be badly out of date. Here, therefore, is a review of current thinking about human evolution.

To understand who human beings are and how they came to be what they are today, primatologists and evolutionary biologists look to humans' closest relatives, other primates. The closer the taxonomic relationship, the more likely that animals will have a similar biology. Genetically, humans' closest living relatives are the two chimpanzee species, the common chimpanzee and the bonobo (once known as the pygmy chimpanzee). Humans are only slightly more distantly related to gorillas. But contrary to the common understanding, humans are not descended from chimpanzees, or even from a chimpanzee-like ancestor. For example, the last common human-chimpanzee-gorilla ancestor was not a knuckle-walker, as is commonly pictured. Biomechanical analyses of the shoulders, elbows, hands, vertebral column, hips, knees, and feet show that human ancestors never walked on all fours; humans evolved from an upright-positioned arboreal climber that walked on two legs when on the ground, as gibbons and orangutans do today (Hart & Sussman, 2009; Lovejoy Lovejoy, Suwa, Simpson ,Matternes , & White, 2009; Schmitt,

2003; Thorpe, Holder, & Crompton, 2007; White et al., 2009). There are, in fact, as many or more biological, behavioral, and physiological parallels between gibbons and humans as there are between chimpanzees and humans (see figure 1).

Figure 1. Branching time of hominoid and Old World monkey lineages.

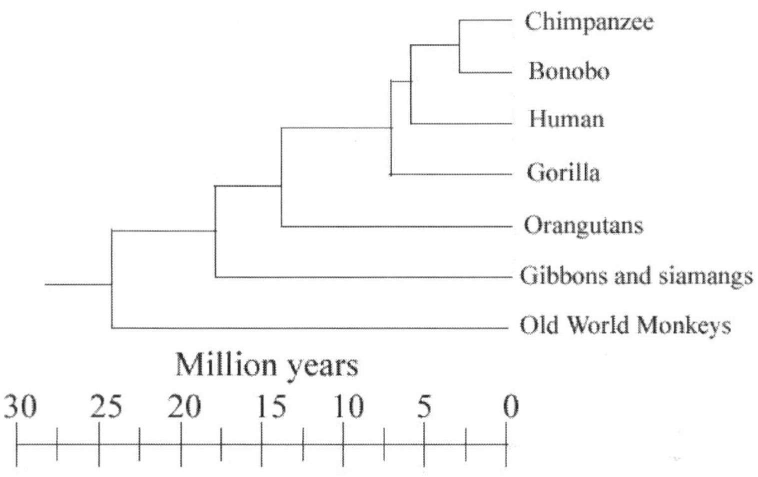

Source. Adapted from Goodman, Grossman, & Wildman, 2005. Courtesy of Lawrence Grossman.
Note. The most primitive possible primate fossil (*Purgatorius unio*) was dated to 64 million years ago. New World monkeys split from Old World monkeys about 40 million years ago.

Humans and chimpanzees are so closely related, with a 98.9% similarity between their DNA[1] (Goodman et al., 2005), that for a short while, some taxonomists suggested that they be classified in the same genus, renaming chimpanzees as *Homo troglodytes* (Diamond, 1992; Goodman, Tagle, Fitch, Bailey, Czelusniak, Koop, & Benson, 1990; Watson Easteal, & Penny, 2001). Partly because of Jane Goodall's research that had found that her provisioned (artificially fed) chimpanzees were capable of infant killing, cannibalism, and murder, the popular press and various primatologists leaped to the conclusion that humans are naturally barbarous, just like their close relative the chimpanzees (Goodall, 1986) (see chapter 3).

However, chimpanzees and human beings are very different in ways that count: in behavior and especially in relationships between males and females (Diamond, 1997). Research in anthropology, physiology, and anatomy as well as other evidence shows that since diverging from a common ancestor about 5.4 million years ago (see figure 1), human forebears and chimpanzees evolved completely different mating systems. Not only that, but chimpanzees and bonobos, which are more similar genetically than humans and chimpanzees, also evolved different behaviors. The mating system of bonobos is no more like that of chimpanzees than humans' is, and bonobos are in many ways more similar to humans than they are to their nearest kin, chimpanzees. Jared Diamond argued on multiple lines of evidence that the common gorilla-chimp-human ancestor lived in harems (one male in a stable mating relationship with several females). Gorillas maintained this model and chimps and bonobos invented promiscuity (though with very different expressions of it), but humans lived in nuclear families. Nuclear families and their origins are explored in chapter 6.

MILESTONES IN HUMAN EVOLUTION

- Gibbon-human last common ancestor, 14.9 million years ago (MYA)
- Orangutan-human last common ancestor, 11.3 MYA
- Gorilla-Human last common ancestor, 6.4 MYA
- Chimpanzee/bonobo-human last common ancestor, 5.4 MYA
- *Ardipithecus ramidus* (nickname "Ardi"), lived 4.4 MYA
- *Australopithecus afarensis* (one nicknamed "Lucy," 3.2 MYA), lived 3.8 to 3.0 MYA
- *Australopithecus sediba*, lived in South Africa 1.977 MYA and may have been a human ancestor

- *Homo habilis* (Latin for "handyman," referring to its tool-making ability), lived in Africa 2.5 to 1.8 MYA
- *Homo erectus/ergaster* (Latin for "upright human"), evolved 1.8 MYA in Africa and persisted to about 500,000 years ago in Europe and 40,000 years ago in Java (*H. erectus* reached Java and China 1.0 MYA)
- *H. neandertalensis* (nickname "Neanderthal"), evolved in Europe and lived 200,000 to 30,000 years ago
- *H. sapiens* (Latin for "wise human"), evolved in Africa between 200,000 and 150,000 years ago

*Other pre-*Homo* and *Homo* species are not included in this list.

Sources. Goodman et al., 1998, 2005; *National Geographic*, February 1997, July 2010; Pickering et al., 2011; Pollard, 2009; Stauffer, Walker, et al., 2001; Tattersall & Mearns, 2000; White et al., 2009.

Humans are unique. They are apes but not chimpanzees. Little can be learned about human behavior from the study of chimpanzees. To understand human beings, one must go further back, to the ancestor of all apes, including the gibbons, so-called lesser apes. All apes lack tails, have mobile shoulders, and can stand erect. Climbing and feeding in an upright posture (sitting or hanging) and walking along branches using long arms for support define the ancestral locomotion common to all apes (Chivers, 1984). An interesting feature shared among all hominoids (apes and humans), but absent in monkeys and other primates, is a ligament in the shoulder, the coraco-acromial ligament, and associated projections of the shoulder joint, that permits unique locomotor and feeding motions (Ciochon & Corruccini, 1977). This shoulder flexibility combined with leverage and power allows a chimpanzee to swing a stick over its head in a threat display, and a quarterback to accurately throw a football into the end zone from midfield.

According to archeological findings, humans' very early forebears are thought to include *Ardipithecus ramidus*, which lived in both terrestrial and arboreal environments 4.4 million years ago (MYA), and *Australopithecus afarensis* (3.8 to 3.0 MYA), which was entirely terrestrial. Recently, bones from *Australopithecus sediba* (1.977 MYA) from South Africa were found that seem especially likely to belong to a *Homo* ancestor (Pickering et al., 2011). This simplified sketch omits some prehominids that may or may not lie in humans' direct lineage; others will no doubt be identified as more bones are found, making the family tree look more like a bush (see figure 2). Some proto-hominids may have had much different mating systems and lifestyles.

Hominids evolved in Africa, from where at least three waves of species spread out over much of Europe and Asia. First was *Homo habilis* (literally "handyman"), the first tool-maker, who emerged in Africa around 2.5 million years ago. About 700,000 years went by before the appearance of *Homo erectus*, a tall, gracefully slender protohuman looking much like modern human beings. Its name in Latin means "upright human," but we prefer to call it "the Wanderer." The first great humanoid traveler, *H. erectus* reached Java around a million years ago and China soon after.

Back in Africa, *H. erectus* evolved into *H. ergaster* (some anthropologists consider *ergaster* merely a later form of *erectus*, not a full species. *H. ergaster* is sometimes equated with *H. erectus*). *H. ergaster* gave rise to Neanderthals (*H. neandertalensis)*, whose ancestors reached Europe and the Middle East around 350,000 years ago and further evolved there, reaching their final form about 130,000 years ago. *H. ergaster* was also the African ancestor of modern humans, *H. sapiens*, who reached Australia via India at least 50,000 years ago, replacing most *H. erectus* along the way. At least two intermediate species, *H. antecessor* and *H. heidelbergensis*, preceded *H. sapiens* in Europe.

Figure 2. Branching pattern of the human family tree.

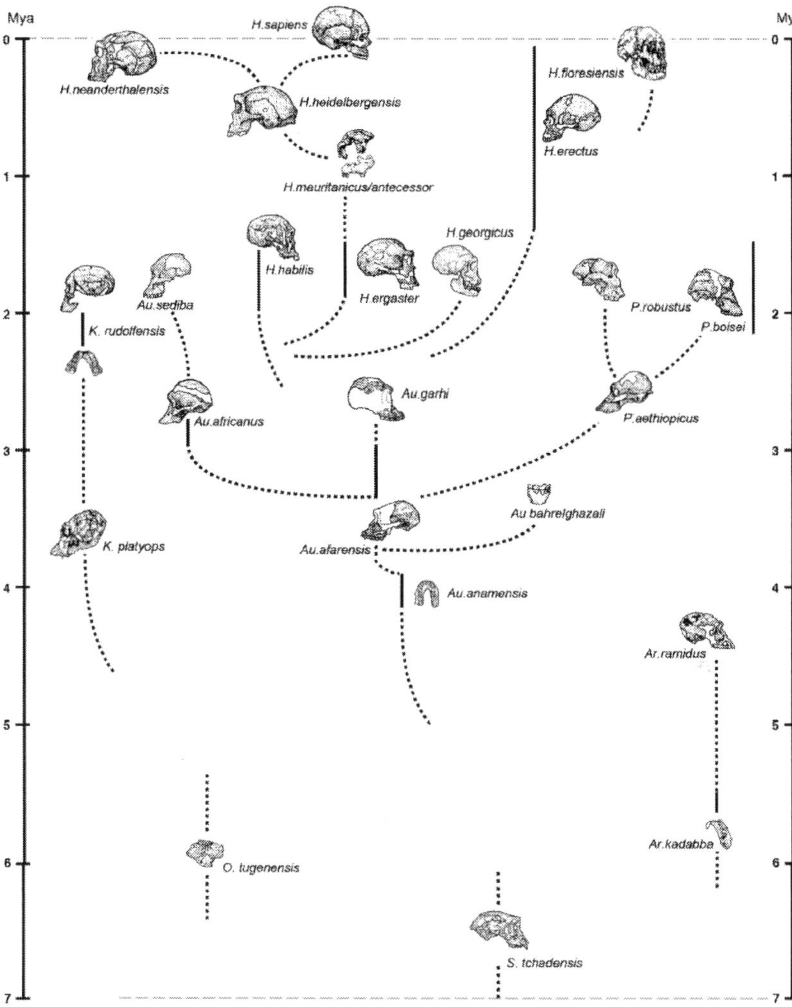

Source. Tattersall, 2010. Courtesy of Ian Tattersall.

In another wave of emigration from Africa, modern humans entered Europe and the Middle East about 46,000 years ago and began replacing

Neanderthals 30,000 to 20,000 years ago. Gibbons (2011) reported that Neanderthals and modern humans' ancestors interbred, because today about 2.5% of the DNA of people from Europe and Asia comes from this species, but Manica (2012) showed that the trace of Neanderthal DNA in non-African *Homo sapiens* more likely resulted from the two species having a shared common ancestor, rather than from hybridization.

Scientists distinguish physically "fully modern humans," *Homo sapiens sapiens*, dating from around 150,000 to 100,000, from "behaviorally modern humans," dating from around 50,000 years ago. It is likely that changes occurred in soft tissues between 100,000 and 50,000 years ago, such as additional complexity of the brain, reduced hairiness, and a reconfigured larynx that facilitated speech (although this may have occurred earlier); such changes do not show up in old bones but are revealed as genetic changes in DNA. Of course, genetic changes continue to mold appearance and behavior.

CULTURAL ACHIEVEMENTS OF PROTOHUMANS AND HUMANS

- Stone tool use and meat eating by scavenging: *Australopithecus afarensis*, 3.4 MYA
- Stone tool manufacture: *Homo habilis* (cutting tools), beginning 2.5–2.6 MYA
- Bone and stone tools: *Homo erectus*, 1.5 to 0.3 MYA
- Hunting: late *Homo erectus* (there is uncertainty about what was hunted and how: probably not big game [Hart & Sussman, 2009])
- Speech: *Homo neandertalensis* and *Homo sapiens*
- Systematic use of fire: *Homo erectus* in Africa, 1.0 MYA; *H. neandertalensis* in Europe, 400,000 YAGrammatical language: estimates range from 300,000 to 50,000 YA
- Scraped hides as clothing: about 300,000 YA
- Sewn, well-fitting clothing: around 20,000 YA

- Stone-tipped projectiles (spears, spear-thrower darts): 40,000 YA in Africa and Europe
- Murder by and of Neanderthals: Sima de los Huesos cave Spain containing at least 27 skeletons dating to c. 350,000 BC, some showing trauma may have been caused by crude weapons, although none were found.
- Murder of Neanderthals: 47,000 YA in Iran; 36,000 YA in France. Stone point injuries may have been inflicted by either Neanderthals or *Homo sapiens*. Some paleontologists believe that *Homo sapiens* exterminated both *Homo neandertalensis* and *Homo erectus*. This would not be "murder" but the moral equivalent of modern medical experiments on primates.
- Murder: *Homo sapiens* killing *Homo sapiens*, 20,000 YA
- Early transition to agriculture: 19,000 to 11,500 YA east of the Mediterranean Sea
- Bow and arrow: 12,000 YA
- Mass murder: 12,000 YA (a grave site in Sudan)
- Domestication of wheat: 12,000 YA east of the Mediterranean Sea
- Genocide: about 3300 YA (Bible, Numbers 21:35: "So they [Israel] smote him [Sihon, king of the Amorites], and his sons, and all his people, until there was none left alive and they possessed his land.")

*MYA: million years ago. YA: years ago.

Sources. Ambrose, 2001; Hart & Sussman, 2009; McPherron, Alemseged, Marean, Wynn, Reed, Geraads, & Bobe, 2010; Provan, Long, & Longman, 2003; Rantala, 2007; Roebroeks & Villa, 2011; Shea, 2006; Thorpe, 2011.

Development of Nuclear Families

We discuss nuclear families in chapter 6, but a word on their presence in humans is needed here. Nuclear families are important to the discussion of aggression because in multimale-multifemale mating systems (groups comprising a number of males and females), males may have to fight with other males for access to females or forgo mating entirely. In species with nuclear families, such as gibbons and humans, there is far less need for aggression once pair bonds are established. We chose to be sparing with the term "monogamy" because it is value laden and ambiguous in certain contexts. Nuclear families are usually characterized by a mated adult male and adult female in a stable pair bond and their dependent young.

The differences between males and females of a species reflect different mating systems. In primate species in which males compete with each other for mating rights, the males are larger than the females (e.g., gorilla males are 60% larger than females) and have larger canine teeth (larger than those of the females as well as those of less combative species) (Leutenegger & Kelly, 1977). In these species, females usually have visual and other advertisements of estrus: Rump swellings and color changes advertise their receptivity, and it is only (or mainly) these receptive females that the males fight over. By contrast, in mating systems that obviate the need for constant male combat, the size difference between males and females is minimal, the canine teeth are hardly larger than the other teeth (and are no larger in males than in females), and the females show little outward sign that they are ovulating. This describes humans[2] and gibbons, except that both male and female gibbons have equally large, sharp canines that they use to defend family territories against other gibbons. Mated gibbons never use these weapons on each other or on their offspring. That the fossils of the earliest protohuman ancestors do not have large canines[3] suggests that they had little need for combat with others of their species. Because this was some 2 million years before tool use, weapons could not have been substituted for large canines.

What is the case for more recent protohuman ancestors? In human evolution, anthropologists used to assume that although a moderate reduction in canine teeth and body size dimorphism occurred in *Homo habilis* and continued in *H. erectus*, the major reduction was in archaic *H. sapiens* and *H. neandertalensis*, signaling a shift to a less polygynous mating system (Leutenegger & Kelly, 1977). However, the antiquity of hominid nuclear families was pushed much further back in time in 2003 with the discoveries of many more skeletons of Lucy's species (*Australopithecus afarensis*). They showed that contrary to the conclusions based on the first few specimens, the sexual size difference between males and females was similar to that of modern humans. The scientists reporting this concluded that nuclear families and the associated male parental care of infants can be traced at least as far back as Lucy, 3.8 to 3.0 million years ago (Reno et al., 2003).

In addition, unlike other modern apes, australopithecines were completely bipedal. New foot bones available for study revealed an arch: Lucy had a spring in her step. She could run (Ward et al., 2011). Now able to roam far and wide on the savannah, human ancestors were no longer restricted to forests. This also points to nuclear families: With thin populations spread across a vast landscape, a male not associated with a female could not be sure of finding a mate when she was ready to breed, would not be there to ensure his progeny's survival, and would not be sure that the offspring were his—a prerequisite for paternal investment. Staying with her solved all three problems.

A new protohuman species discovered in 1994 but not fully described until 2009 further upset conventional wisdom. *Ardipithecus ramidus*, or "Ardi," lived 4.4 million years ago, preceding Lucy by more than a million years. Ardi shows both arboreal and terrestrial adaptations. Although Ardi walked upright when on the ground, it was still comfortable in trees because its large toe diverged from the others, reflecting the foot-grasping ability of its arboreal ancestors. Crucially, its hand, wrist, and other bones showed no ability at all for knuckle-walking, proving that modern humans' ancestors used upright postures in the trees and retained them when they became more terrestrial (Lovejoy, Simpson, White, Asfaw, & Suwa, 2009; Lovejoy, Suwa, Simpson, Matternes, & White, 2009). Ardi had small canine teeth that were the same size in both males and females, with a "less threatening diamond shape" compared to the sharper, pointed shape of previous extinct and other living apes (Suwa, Kono, Simpson, Asfaw, Lovejoy, & White, 2009, p. 69). The teeth, together with the lack of body size difference between males and females, led to the conclusion that there was little male-male competition and "a tendency for male-female co-dominance"(Suwa et al., 2009, p. 99).

Many scholars believe that pair bonding was behind increasing brain size. The shift to pair bonding would have given all males a chance of reproduction and hence a stronger interest in the group's welfare, making human societies larger and more cohesive.[4] Or, perhaps there were fission-fusion groups, as in today's chimpanzees. Either way, greater cognitive ability was needed to keep track of, and manipulate, relationships in societies of increasing group size (van Schaik & Dunbar, 1990). For primate species, there is a strong correlation between the number of individuals in an average group and the size of the neocortex of these individuals; presumably the increased brain size is necessary for the animals to keep track of the greater number of possible social interactions among group members (Kudo & Dunbar, 2001). As in modern hunter-gatherers, such as the !Kung of southern Africa (Shostak, 1981), group size probably fluctuated as families converged and dispersed seasonally, making two or more levels of organization, such as family, clan, and tribe.

By 1.8 to 0.5 million years ago, *Homo erectus*, the last hominid species preceding modern humans, was nomadic and lived in nuclear families. The Wanderer was built for walking and running (Bramble & Lieberman, 2004; Dunbar, 2003; Steudel-Numbers, 2006): It was as tall as modern humans and long-legged, with sturdy knees; limber, strong feet with flexed arches to absorb impact; and narrow hips shaped for an efficient stride.[5] By then, females had lost the visible signs of estrus, such as the prominent rump swellings that other apes have, and had evolved the ability and willingness to mate at any time (though not necessarily with anyone).

The Wanderer families were not isolated but were members of communities—clans, possibly—of around 100 to 120 individuals, which is 2 to 3 times the group size of Lucy's genus, the australopithecines. Females, being twice the size of australopithecine females, had higher energy needs during pregnancy and lactation (Aiello & Key, 2002; Roth & Dicke, 2005). Males had little need to fight each other, except probably during adolescence when nuclear families were being formed. Its brain being about 1,100 milliliters (30% larger than that of its predecessor, "Handyman" [Rightmire, 1993]), the narrow pelvis of the female meant that much of the brain growth had to occur after birth, making offspring dependent on their parents for much longer (Johnson & Earle, 2000). This implies the need for one or both parents to carry infants as they traveled, at least until the advent of settlements late in the history of modern humans.

For the Wanderers, delayed maturity, or neoteny (also discussed in chapter 6), meant that to ensure succession, males had to invest in serious, prolonged care of their progeny. Nuclear families often included infants, toddlers, and adolescents. Not only did the adults have to provide shelter and sustenance, but when whole families traveled together—as must have been common in order for them to reach Java by a million years ago—the men had to carry a toddler and perhaps lead another by the hand while their mates carried infants (Ellis, 2003; Joffe, 1997; Kaplan, Hill, Lancaster, & Hurtado, 2000). Bear in mind, this was not just walking across savannah, but clambering down and up mountain cliffs, wading through swamps, and crossing rivers. Increasing longevity meant that elders were often present in nuclear families. They contributed to the group's survival by helping with the kids, finding food, and imparting knowledge that could be gained only through longevity, such as how previous generations survived rare environmental catastrophes (Dagg, 2009; Ellis, 2003; O'Connell, Hawkes, & Blurton Jones, 1999; Roth & Dicke, 2005). These are also features of modern humans which persist in hunter-gathering groups today (Marlowe, 2003). This was the social platform that launched humanity.

The archaeological discoveries of the last decade or so have fundamentally altered the view of the evolving role of humans' male ancestors over at least the last 4.4 million years. Rather than battling each other over access to females and either forcing females to have sex or paying for it (i.e., the "food for sex" scenario, which we discuss in chapter 6), the evidence now shows that men were peaceable, often pair-bonded, relatively equal partners in long-term nuclear families.

Chimpanzees diverged from humans' line; humans did not diverge from theirs. They are the ones that failed to stand upright, never left the jungle, and never cooperated fully with each other and whose males never learned to protect females and offspring. Men will fight, even to the death, to protect their families, but humans prefer to live in peace—as most do most of the time (Fry, 2006). Nothing gives humans greater pleasure than, together with a mate, watching their children grow strong and wise and provide in turn grandchildren to teach and spoil. This is what it means to be human.

HORMONES

The fossil record shows when and how humans evolved, but not why. A human being can throw a spear because humans' shoulders, preadapted to hanging from branches of trees, allow the arms to rotate nearly 360 degrees. Modern humans' ancestors could invent bows, arrows, and spears because their brains were large and complex enough to construct hunting weapons, as well as maintain social relationships and speak languages. Humans had these abilities for hundreds of millennia but did not begin to use them to attack each other until around 20,000 years ago. This change in behavior involved ancient hormones tuned to new uses, learning, and culture.

The human endocrine system is well known and so similar to those of other primates, and indeed of other mammals in general, that scientists in laboratories use animals to research the hormonal basis of human behavior. Until the last 15 years or so, hormones from animals in the wild could be analyzed only from blood samples, which made their study extremely difficult. Because of advancements in chemical analysis, hormone levels in biological samples can now be obtained noninvasively—from feces, urine, saliva, and hair (Anestis, 2010).

We will omit here a discussion of the hormones involved in growth and reproduction, even though they are well known to influence behavior as well as biology, and focus instead on those involved in stress and aggression. One of these hormones is testosterone, which produces the secondary sexual physical and behavioral characteristics of males. Testosterone is often equated with male aggression, but this is not exactly the case. In fact, the reverse can be true: In one study, men with missing or small testes that did not produce much or any testosterone were found to have unusually high self- and partner-reported levels of tension, anger, and fatigue that they often expressed as aggression and violence. Experimentally, Daryl O'Connor and his colleagues (O'Connor, Archer, Hair, & Wu ,2002) found that these men had a low ability to control their behavior when required to do so in certain social situations; administering testosterone led to better control, less tension and anger, and a general increase in vigor.

Testosterone does drive certain nonreproductive male behaviors, but the link with aggression and violence per se is mainly in certain subpopulations, such as prison inmates and alcoholics (Campbell, Dreber, Apicella, Eisenberg, Gray, Little, & Garcia, 2010). Behavioral scientist Kaj Björkqvist (1994, p. 177) stated that "it is incorrect, or rather, nonsensical, to claim that males are more aggressive than females." His and similar theories, that the genders have about the same level of aggression but express it differently, depend on the definitions of aggression and violence. In the general male population, high testosterone levels along with other hormones (serotonin, dopamine) are associated with impulsiveness and thrill-seeking (Campbell et al., 2010).

Here lies the explanation for the behavior of the three youths who were introduced in the beginning of this chapter. As decades of actuarial data on car accidents prove, young males are risk-takers. Testosterone levels peak in the early twenties and decline thereafter. But in the described case, culture was also involved: The youths had not yet learned the boundaries of acceptable behavior, for which the neighborhood men provided a lesson. It is easy to imagine that high risk-taking, sensation-seeking, and impulsive behaviors would be an evolutionary advantage for men. A little extra testosterone may have primed humans' ancestors to defend against predators and hunt dangerous game, just as it may help today's teenage boy to get his nerve up to ask a girl to a dance. But this is speculation.

What is not speculative is that testosterone's influence on impulsivity and thrill-seeking works by lessening social inhibition. This is where hormones and culture interact. At least once in their lives, if they are to reproduce, men need to probe women to see who might be receptive to courting, and they need to find the socially correct situations for this and to use socially correct approaches. Women also have this need, but because they have a biological imperative to be more selective, they learn early not to send the wrong cues and to dissuade men's approaches when the cues have been misread. Unfortunately, sending and reading these signals is imperfect at best. Cultures vary, even when biology does not. Different interests are involved; if these are compounded by either party's excessive impulsiveness or reticence, there are countless ways for an interaction to go awry.

Besides misreading social cues, there is also a biological basis for some sexual misconduct. People who have a variant of the DRD4 gene may produce excess dopamine, a natural opiate (Campbell et al., 2010; Dreber, Rand, Garcia, Wernerfelt, Lum, & Zeckhauser, 2010; Eisenberg, Campbell, MacKillop, Modi, Dang, Lum, & Wilson, 2007; Eisenberg, Apicella, Campbell, Dreber, Garcia, & Lum, 2009). Justin Garcia, a DRD4 researcher, noted, "The DRD4 gene, which affects the dopamine receptors in the brain, is important for pleasure and reward. It goes off when you jump out of an airplane, engage in sexual behavior, drink alcohol and gamble" (interviewed by Ryan, 2010, p. A9). Like genes that govern height, certain variants or mutations will affect how the so-called infidelity gene expresses itself.

Complementing aggressive hormones are those that encourage friendly, or affiliative, behaviors. Paternal care of infants is associated with nuclear families in primates and with the hormones vasopressin and possibly prolactin and testosterone (Fernandez-Duque, Valeggia, & Mendoza, 2009).

Men's testosterone normally falls when they get married or otherwise partnered and falls again when their babies are born, which scientists interpret as a natural switch from mate-seeking to parenting (Campbell et al., 2010; Gettler, Mcdade, & Kuzawa., 2011aGettler, McDade, Feranil, & Kuzawa, 2011b; Kuzawa, Gettler, Muller, McDade, & Feranil, 2009). At the same time, other hormones that are associated with love, trust, contentment, and satisfaction in relationships increase. Evolutionary biologists view these normal changes as associated with the pair bonding and paternal care of young that developed millions of years ago in protohuman ancestors (Gettler, 2010). Oxytocin promotes affiliative (trusting) behaviors, lowering the natural resistance that animals have to the too-close proximity of others (Kosfeld, Heinrichs, Zak, Fischbacher, & Fehr, 2005). Endorphins such as vasopressin and prolactin, released in the brains of husbands and fathers, make humans no less than mice and monkeys feel pleasure or euphoria (Fernandez-Duque, 2007). The brain releases phenylethylamine, a type of endorphin, when a person or a monkey experiences attraction to another (Edgar, 2010). This is a vast topic summarized briefly by Flynn (2011, p. 16): "The neuroendocrine mechanisms that underpin human paternal and grandparental psychology are not well studied, but likely involve the common mammalian affiliative hormones oxytocin and arginine vasopressin, with additional influence from the hypothalamic-pituitary-gonadal and hypothalamic-pituitary-adrenal systems."

It is the tug-of-war between affiliative and aggressive behaviors that determines when aggression becomes too much or is expressed in the wrong situations. These examples are necessarily oversimplified. Biologists talk of "signaling pathways" that may involve a long chain of biochemical causes and effects between a stimulus and a response. In the end, these compounds work by either stimulating or inhibiting physiological responses and behavior.

Although hormones for flight-or-fight, trust, and attraction are fundamental and ancient, there are newer, inherited changes that modify the context in which these basic imperatives operate. There are also uninherited cultural traits. That they are not inherited makes them no less fundamental and "right" to the human species' behavior. It is not a question of nurture versus nature: It is both, all the time.

Human male bonding through shared activities and struggles—as hunting and territory defense often were, and still are—was reinforced by hormones, so that males could feel good upon seeing a group of friends come over the hill and could fear seeing strangers. This is not guesswork; it is known for certain, based on the knowledge of hormones and their effects. Similarly, females got along well with each other as well as with the men that they knew, but retained a justifiable apprehension (or excitement in some circumstances, or both) upon seeing men they did not know.

Evolutionary biologists sometimes talk of sexual roles as simple economic exchanges: The man trades food and security and the woman trades sex. (As an indication of how common this view is, Harding Googled "Food for sex" + "archaeology" and got 52,800 hits.) But these are only some of the evolutionarily relevant outcomes. The most important is the survival of offspring. Surely a comely young *Homo erectus* female, seeing a lithe male laugh and smile, carve things for her, and linger with her, would have felt no different than a young woman today in an analogous situation. They may have mated, formed a stable pair pond (not necessarily in that order), raised a family, and continued to admire each other in their senior years, as most *Homo sapiens* couples do today. Humans' forebears traded, and people still trade more than food, security, and sex. Mates share trust, love, tenderness, compassion, excitement, and, yes, jealousy and anger. This is known because modern humans have the same hormones, producing the same effects, as the first protohumans had some 5 million years ago.

HAIR

It may not have escaped one's notice that humans, uniquely among primates—indeed, among virtually all terrestrial mammals, that is, excluding the marine and fossorial (underground living) ones—are naked. Or that the naked human form is, well, exciting to humans. These two facts mark an important milestone in human evolution.

Hair density thins in hominoids as one proceeds up the family tree from gibbons, with the densest hair, to humans, with the thinnest; orangutans, gorillas, and chimpanzees are intermediate (Schultz, 1933). All primates, except humans, groom each other, that is, they pick out parasites and bits of debris from each other's hair. The practice is so ubiquitous among primates that its absence in humans begged for an explanation, and Robin Dunbar (1996) provided it: grooming was not merely hygienic but an important element in the intensely social milieu of primates, that most social of mammalian orders. So far, Dunbar merely echoed the standard paradigm among biologists, but then he went a big step further: when (or possibly even because) humans developed the use of language, they had no further need for social grooming.

But humans still needed to groom for hygiene and health, right? No, not if humans did not have any hair. Hermelita Winter and her colleagues (Winter, Langbein, Krawczak, Cooper, Jave-Suarez, Rogers, Praetzel, et al. 2001) analysed the DNA of the human pseudogene ψhHaA for hair kerotin, which is inactive in humans, and found functional analogues in the gorilla and chimpanzee. Based on their calculations, they concluded that the human one was deactivated around 200,000 years ago, making humans naked. This was, they said, long after *Homo ergaster* had become extinct and about the time that the human line acquired its contemporary appearance. This is consistent with other evidence that humans began using scraped hides as clothing about 300,000 years ago, and began sewing well-fitting clothing 20,000 years ago (Rantala 2007). Why did humans begin to use clothing?

Pagel and Bodner (2003) proposed that humans evolved hairlessness to reduce parasite loads, and this fit with Dunbar's theory: if humans no longer needed social grooming, why not dispense with health and hygiene, too? Pagel and Bodner (2003, p. S117) went further and suggested that:

> ...hairlessness is maintained by these naturally selected benefits and by sexual selection operating on both sexes. Hairlessness is made possible in humans owing to their unique abilities to regulate their environment via fire, shelter and clothing. Clothes and shelters allow a more flexible response to the external environment than a permanent layer of fur and can be changed or cleaned if infested with parasites.

The sexual selection idea proposes that less hairy bodies were initially pleasing to mates, who could thereby tell if a prospective mate had a disagreeable parasite load. Eventually, nakedness became pleasing, period. The gender difference in amount and distribution of body hair further supports the sexual selection hypothesis (Rantala 2007). Rantala (2007, p. 3), elaborated that "as human society depends upon the specific 'bonding' of pairs of individuals, anything adding to the pleasure of contact may have conferred a biological advantage."

Other researchers, such as Toups, Kitchen, Light & Reed (2011), used a different approach. Human ancestors, like other apes, once had one species of louse but modern humans have three: one that lives in the pubic hair, one that lives in clothes, and one in the head hair. By analyzing DNA to determine divergence times of these lice, Toups's team proved that humans began wearing clothes between 170,000 and 83,000 years ago. By inference, this is when we began taking off our clothes, in the right situations.

Genes

"There is no such thing as an aggression gene," according to Han Brunner, a neuroscientist (Richardson, 1993; Brunner, Nelen, Breakefield, Ropers, & van Oost, 1993). Genes do, however, make human beings who they are, defining not only the species but individuals within societies. They provide the substrate on which biology and culture set the context for people's actions. Like hormones, genes do not show up in the archaeological record. Luckily, their history is written in every cell's DNA. After first being awed by the close genetic match between chimpanzees and humans, researchers have since learned that small changes can make huge differences in appearance, ability, and behavior.

Neuroscientists classify genes in three functional types: structural, operator, and regulatory. Structural genes code for specific proteins that make body parts such as hands, glands, and so on. But these genes are few and far between along the length of a nuclear DNA strand, and they do not vary much. An orangutan's hands and fingernails are not much different from humans', and the genes that make them are identical, or close to it. The operator and regulatory genes control the timing and extent of their expression. Henry Gee (2004) provided a compelling summary of how they work: The three types involved in related functions are frequently close together on the chromosome and operate as a network. Because they are close together and the structural genes are relatively few, a mutation in a length of DNA is more likely to affect the timing and expression of a gene than the gene itself. This is how gibbons, for example, can evolve exceptionally long arms for a primate, but their arms have the same bones, muscles, vessels, and sinews as other primates.

Some 30 genes are known that affect social behavior, mostly identified in nonhuman animals (Wade, 2006). One of them, the DRD4 gene, is associated with testosterone levels and the sensation-seeking behavior discussed earlier (Campbell et al., 2010). Another is a gene that codes for monoamine oxidase-a (MAOA), the rare lack of which produces overly aggressive men of low IQ (Brunner et al., 1993).

Some genes change rapidly, and new mutations can alter physiology and behavior. An example is a mutation, probably in one person, that conferred the ability to digest milk about 7,200 years ago in the Fertile Crescent. That society had already domesticated cows, sheep, and goats for meat and hides. Milk drinking helped this person and his or her family, especially infants, survive because animal milk can supplement a mother's milk. This ready source of food that could travel with the family, besides providing obvious benefits to infants and other family members, would improve the mother's health (lactation is a huge energy drain) and help her to engage in other activities to benefit her family and the community. Modern humans had already occupied Europe for nearly 40,000 years, living mainly as hunter-gatherers. Lactose tolerance and consequent changes in animal husbandry permitted a rapid increase in the population of those who had the new gene. They already had the animal husbandry "package" and quickly spread north and west, replacing—sometimes violently, as at least one mass grave attests—hunter-gatherers (Schulz, 2010). They spread throughout northern European in the space of about 100 generations. In this example, the gene did not cause violence but established the context, an imbalance of power, in which violence helped one group prosper at the expense of another.

BRAINS

As well as their inherited hormones and genes, humans have huge brains designed to develop and manipulate increasingly numerous and complex relationships and to support language.[6] Apart from their large cerebral hemispheres, human brains are so similar to those of other mammals that "if one learns the subcortical neuroanatomy of one mammalian species, one has learned the ground plan for all other mammals" (Panksepp, 1998, p. 60). Human brains have specific regions that light up an MRI when subjects experience pleasure, fear, and trust; these areas "fire" when individuals cooperate (Rilling, 2011). Along with such friendly[7] behaviors that engender cooperation and attachment are agonistic behaviors and the behaviors that are needed to avoid them (see chapter 5). Scientists have found various parts of the brain linked to aggressive behaviors (Johnson, Bodon, & McCroskey, 2008). Like human beings, chimpanzees can use their brains to be deceitful—for example, feigning not to see where a colleague hides food, the better to steal it when he is absent. This ability is a hallmark of intelligence that not many primates have, but because human beings and chimpanzees have it, it is a good guess that their joint forebears 5.4 million years ago had it, too.

Humans have hormones for trust and intelligence for deceit; understanding based on brain power must now be added to behavior based on hormones. With the ability to deceive came the ability to perceive cheating by others (Dunbar, 2003; Johnson & Earle, 2000; Wade, 2006). People learned right from wrong—"right" being defined as what is considered to be so within one's culture. Societies began developing ways to punish cheaters for the good of the community, for example by shunning. Customs developed cultural ways of recognizing right and punishing wrong. This ability is also part of being human. Although customs differ by culture, the need for them and the ability to develop and apply them is universal. Therefore, priority can never be assigned to behavior based solely on hormones over behavior based on knowledge of right and wrong, as some popular commentary would have it. Hormones may move men, deep down in their reptilian brains (a useful metaphor, and not entirely inaccurate), to covet their neighbor's wife, but most decline to act on this inclination because they know right from wrong.

In the early stages of the study of the evolution of behavior, sociobiologists thought that an organism should do only things that increase its own "fitness," taken generally to mean its ability to pass its own genes to offspring (Wilson, 1975). Some still think so (Dawkins, 2006). They had difficulty, however, explaining the evolution of such human values as honesty, generosity, compassion, and altruism. Using animal models, they developed the "kin selection" theory (Foster, Wenseleers, & Ratnieks, 2006; Michod, 1982). It posits that because one's near relatives have some of one's genes—for example, a woman's son has half her genes and her nephew has one quarter of them, inherited from her own mother or father —one can pass on some of one's own genes by ensuring the relatives' survival.

Other research, mainly by mathematical modeling, shows how this idea can extend to a clan or, further, to a tribe. However, it is markedly unsuccessful at explaining why anyone would risk loss of life or property by defending unrelated people, as many people do—not only in war, but in everyday, trivial events like the one that introduced this chapter. It also does not explain why a bystander would run to catch a thief or stop to help an accident victim, or why people feel sad when perceiving others' misery.[8] These values, so very characteristic of the human species, cannot be explained by genes and hormones. And it is difficult to explain how they can be set aside on such a large scale as war (a behavior that emerged late in the human experience and is cultural rather than innate). This is an ongoing field of research of great interest (Bowles & Gintis, 2011; Sussman & Cloninger, 2011).

Endnotes

1. The 98.9% figure, first reported as 98.6% and since revised upward, is only one measure of genetic similarity and difference. For example, it exceeds 99% if one compares only amino acids, and it drops to 86.7% if one compares the sequences, taking into account the multiple insertions and deletions throughout the region of the DNA strand that is being examined (Anzai, Shiina, et al, 2003). For perspective, gorilla DNA orthologs (the coding part) are 98.5% identical to humans', and orangutan DNA is 97.0% identical (reviewed in Goodman et al., 2005).
2. Human males are 15% to 20% larger than females, which is thought to be maintained by sexual selection (i.e., women like slightly bigger men). The larger sexual dimorphism differences, in body size and teeth of other primates, is due to natural selection because, unlike humans, the males must battle each other for access to females.
3. There are other differences than the size of canines. Unlike primates in multimale-multifemale mating systems, the shape of canines in these protohuman ancestors was less daggerlike, and the bottom set did not rub against the top ones to hone them sharp (Suwa et al., 2009).
4. Even today, a woman giving birth who is not supported by a man is more likely to have her infant die than if she were part of a couple. Data from Florida between 1998 and 2005 for more than 1.3 million live births indicate that those babies who did not have a man's signature on their birth certificate were less likely to thrive (Alio et al., 2011). Their mothers were more likely than women with partners to have suffered obstetric complications such as anemia, chronic hypertension, eclampsia, and placental abruption. The infants of these absent fathers had a death rate nearly 4 times greater. The authors of this research estimated that 65% to 75% of excess mortality could have been prevented had the fathers been involved with their partners and babies.
5. Women's hips are not quite as narrow as men's in the upper part that holds the womb, a reproductive compromise that is discussed in the following paragraph, but they are shaped for efficient running the same as men's, especially in the parts that join with the legs and support running muscles.
6. Scientific opinion is divided on the relative importance of language or increasing group size underlying humans' language ability, but both were certainly involved (Coupé, & Hombert, 2005).

7. Scientists use the term "affiliative" for behaviors that facilitate trust and cooperation.
8. A large body of research shows that humans are extremely adept at detecting the emotions of others. Now research has shown that dogs, domesticated at least 15,000 and possibly as much as 100,000 years ago, also have the ability to detect human emotions. This may be unique in nonhuman species (Udell, Dorey, & Wynne, 2011).

Chapter 3

Bias in the Story of Human Evolution

Vicious Apes

According to DNA, human beings are equally closely related to chimpanzees and bonobos (apes) but much less closely related to baboons (monkeys). It has been known for centuries from their anatomy that both apes and monkeys are more similar to human beings than are other species; when Linnaeus worked out his taxonomic relationships in 1758, other apes had not been formally described. He placed human beings and monkeys in the same order, Primates, but this did not sit well with some scholars who sought a separation between "man" and "beast." Linnaeus classified orangutans in 1760 and gibbons in 1771, placing the latter with humans in the genus *Homo*. Chimpanzees were named in 1775. Other scholars separated primates into the orders Bimana (two hands), containing only one species, *Homo sapiens*, and Quadrumana (four hands), containing other apes and monkeys (Lesson, 1827; Martin, 1841).[1] Charles Darwin helped settle the matter in 1859.

This chapter relates in part how male authors over the past decades have chosen to portray the most fierce primate species they were aware of as similar to human ancestors. Was this because they knew more about aggressive species (which were more exciting to watch, write about, and read about than boring ones)? Was it because they supposed that just as men cause massacres and wars, their primate ancestors must have been aggressive, too? Was it because as male authors, they valorized masculinity and aggression in general and their own personal superiority to women in particular?

Apes have a long history of being thought of as stupid and brutal. William Shakespeare, writing *The Tempest* in 1610, described a person's forehead as "villainous low" with reference to apes. Early naturalists were curious about orangutans, in part because the Malay word means "man of the forest"; they wondered if they might be human. In 1808, a French naval officer brought a live orangutan to Paris. Describing it, French naturalist Frédéric Cuvier (Geoffroy Saint-Hilaire and Cuvier,1819, translated by James Rennie, 1838, p. 93) noted that it:

> used its hands as we generally do our own, but with rather less facility, arising from want of experience...In order to defend itself, it bit and struck with its hand...[but] generally speaking, it was gentle and affectionate, and showed a natural propensity to live in society.

These and other early descriptions of the gentle, peaceable nature of orangutans (summarized by Rennie in 1838) did not keep Lesson (1827, p. 32) from writing, admiringly, that "Le pongo est sauvage, feroce et courageux; il se defent, dit-on, avec un baton lorsqu'il est attaque par l'homme" ["the orangutan is wild, ferocious and courageous; he defends himself, they say, with a stick when attacked by man"] .

When European men first met gorillas in Africa in the 1860s and 1870s, the apes were described as vicious killers that rushed forward to attack them. Annoyed male gorillas do make bluff charges, causing no harm, but in the early days there was no way of knowing they were bluffing because they were shot dead before their bluff was called. By boasting back home about the animals' power and fury, hunters built up their own adventurous egos. The apes' supposed aggression also helped trigger a demand from museums, zoos, and circuses for gorilla specimens, dead or alive. The precedent was set that large apes were by nature highly aggressive.

Apes came next into public consciousness in the 1920s, with three ape-related excitements. First was the Scopes trial in 1925, one of the most important cases in the history of the United States. This was the trial of John Scopes, a teacher in Tennessee who taught evolution to his class although it was against state law. Scopes lost the case, but it forced millions of people to think about the place of humanity in the biological world, an issue that had been settled, as far as scientists were concerned, with Darwin's *Origin of Species* 65 years earlier. Were humans shaped in the image of God, as the Bible states? Were all the millions of animals in the world created by God, meaning that none had ever evolved from other species? Or were human beings animals that had evolved from other apes to become the people they are today?

Then there was Robert Yerkes, an American psychologist who published the book *Almost Human* in 1925. He had gathered up the limited information available about captive great apes from a number of sources —in German, Spanish, Russian, and French as well as English—in writing his book, which showcased an opulent private zoo near Havana, Cuba; there, he observed caged chimpanzees, gorillas, and orangutans. About them he wrote that the more a person learns about them, "the more helpful lessons for mankind he discovers in their relations to their world and to one another" (Yerkes, 1925, p. vii). Later, Yerkes founded an ape-breeding colony that still functions today as the Yerkes Primate Research Center in Atlanta.

Also in 1925, interest in baboons was galvanized in England when keepers at the London Zoo set up Monkey Hill to attract visitors. The keepers transferred 100 hamadryas baboons to an island 100 feet long and

60 wide where they could frolic about without being bothered by their human voyeurs. However, only six of the baboons were females, so there were baboon quarrels about which males should mate with them. Later keepers added a few more males and 30 more females. "That should solve the problem," they probably said to each other with satisfaction as they surveyed the swarm of animals covering the island. It did not. Zoologists then knew nothing about baboon behavior in the wild—in this species, each adult male is accompanied by one or two females that are always with him—and mayhem continued. Far from being an amusing and amazing exhibit, it became a horrific one: Onlookers were aghast to see many animals with broken legs or bites and bloody gashes on their bodies. Males in their frenzy to get females for themselves ripped them up, crushing or strangling their young in the struggle. Of the 15 infants known to have been born into the group, only one survived. Thirty females died of their injuries, as did 62 males. More than two thirds of the baboon population was wiped out. Monkey Hill was indeed an amazing exhibit, but not the one the zoo had intended.

Solly Zuckerman, a zoo man who had made notes of the behavior of these stressed baboons, wrote a book about them and other primates called *The Social Life of Monkeys and Apes* (1932), which widely popularized the idea that large monkeys such as baboons were killers.[2] Given the name of his book, Zuckerman could have had no idea how unrepresentative the zoo baboons were among the 200 or so species of primates. He also did not realize that animals in captivity do not behave like those in the wild. How could they, cooped up with individuals they might hate and eating food they did not choose? His conclusion that male baboons were natural killers was unfounded because it was based on animals that were crowded far too closely together, with no way to escape, and in an unnatural sex ratio.[3]

Zuckerman linked baboons to human beings by using words to describe them such as "overlord," "bigamous family party," "harem," "bachelor" (meaning a celibate male), and "prostitution" (Dagg, 1983). No wonder readers thought that people, or at least men, must be aggressive, too, like their primate baboon-like ancestor. If human males were aggressive because of their genetic inheritance, then this trait was with them forever. This implied that women must put up with men's violence; a woman would be wise to choose and marry a husband who could protect her from other men. Zuckerman, who was knighted for his work, giving it even greater credibility, remained intransigent to the end. As an old man at a primate conference in 1962, he listened to Jane Goodall's findings about amiable wild chimpanzees with disdain, announcing that all he had discovered earlier about killer primates was still true and that Goodall's results were mere anecdotes (Peterson, 2006).

In 1933 came the film *King Kong*, where moviegoers could see the scary gorilla as if in real life. He frightened the wits out of Fay Wray (although any liaison between them would seem difficult given that Kong was depicted as huge enough to grasp her in one hand). This film reflected European culture, in which women were second-class citizens, seldom allowed to be anything but unpaid homemakers or low-paid teachers and nurses. It was men who claimed the power, intellect, and aggression to earn their central place in society, and male animals, like men, were assumed to be dominant in the natural world as well.

An odd experiment set up in the 1950s gives a good idea of society's paradigm that males are dominant. Michael Chance (1956) studied the behavior of rhesus macaques at the London Zoo. Because there were many monkeys leaping about in their enclosure, he designated only the few adult males, which made his observations much easier. They became D1, D2, D3, and so on—*D* standing for "dominant." The females and young were not given names or numbers but were mentioned anonymously only if they were close to one of the males. Needless to say, Chance concluded that life in rhesus communities revolved around the large, dominant males. What other conclusion could there be when the research focused entirely on them?

HUNTING APES AND THE ELEVATION/WORSHIP OF THE BABOON

Zuckerman has a lot to answer for. His book detailing the fierceness of male hamadryas baboons was so popular that readers bought into the idea of male aggression of primates, and therefore of men. In 1959, Sherwood Washburn and his student Irven DeVore from the University of California were anxious to document the "subhuman" behavior of baboons in Nairobi National Park, which, they presumed, would include hunting by the males and male dominance. Because DeVore expected to see male dominance, this is what he observed in his short study of the wild animals. Frustrated because the baboons tolerated his wife being near them but not himself, he started feeding a troop peanuts, causing nasty competition—with the big males aggressively pushing aside smaller males and females as they snatched up the free food (Jahme, 2001). It was clear to DeVore as he watched that the large males were indeed dominant, a condition mirrored in human beings. It did not hurt the comparison that baboons also live a nomadic life on savannahs, just as had protohominids millions of years earlier. Satisfied that their preconceptions had been shown to be true, in 1961 the two men wrote a paper comparing the close social behavior of baboons and "early man" without even taking combativeness into consideration (Washburn & DeVore, 1961).[4]

However, aggression is exciting. It sells books. Best-selling male authors began to write about humans' early ancestors being aggressive too, like baboons, hunting apes that gave rise to the warlike *Homo sapiens* of today.[5] Science writer Robert Ardrey noted in his book *African Genesis* (1961) that although "physically the ape is a little more closely related than the monkey to our human line, still the baboon is the most significant of all the primates" (p. 71). Knowing how to survive life on the ground rather than among trees would have been the same for both baboons and human ancestors, he said.

Ardrey ignored the fact that baboons are monkeys, not apes, and branched off from the human-ape lineage way back in the Miocene, 22 million years ago (Goodman et al., 1998). This was like trying to infer lions' behavior from that of coyotes. At that time, because little was known about the behavior of chimpanzees and other apes, they were ignored. Ardrey insisted that modern humans' predecessors had not been weak vegetarians like other apes and monkeys, but mighty hunters. Human beings were obviously primates, given their anatomy, but unlike all other monkeys and apes, they had far surpassed their primitive relatives because of their ancestors' diet of meat. This was not only because these individuals ate meat—any scavenger could do that—but also because they were hunters who killed the animals that sustained them. Hunting by packs involved careful planning, strategizing about methods, searching out quarry, and eventually dispersing to inhabit new terrain, all activities that encouraged the evolution of larger brains.

Ardrey's evidence for hunting came in part from fossils in the basement of scientist Raymond Dart, who lived in South Africa. Dart had collected ancient fossil evidence of australopithecines along with about 50 baboon skeletons from a cave in northern Transvaal. Many of the baboon jaw bones carried a distinctive mark of trauma, a double-depression scar indicating what had apparently killed them. These indentations looked like they had been inflicted by the end of a humerus arm bone wielded by humans' ancient ancestral males, Dart and Ardrey thought. Ardrey claimed that hunting had set humans' progenitors free: free of the dark forests, free to use their hands in a variety of ways, free from moving on four legs, free from having to eat greenery all day long, and free to use their wits to conquer their new way of living. In a later book *The Hunting Hypothesis* (1976, p. 10), he insisted that "man is man, not a chimpanzee, because for millions upon millions of evolving years we killed for a living" (p. 10): Throughout human evolution, natural selection refined protohuman species for an increasingly predatory life.

But a number of bones with dents apparently made by protohumans (later determined not to be so at all) cannot depict the sweep of human prehistory. Some of humans' early ancestors may have killed infants or sick baboons. A few primates occasionally eat meat, too. But australopithecines were small, around 4 feet tall. Males had small canine teeth completely unlike the large canines of baboons. If they ate much meat, they must have scavenged it. They could certainly not have been mighty hunters, despite the odd evidence that Ardrey set forth.

Ardrey portrayed female australopithecines not as hunters, too, but as a drag on male activity. They had to be based in a stable camp where they could raise their young and care for the sick or injured. The males had to set up a territory around this base camp to protect them, and in case the females should stray when most of the males were away hunting, pair bonds evolved so that a female would remain true sexually to her mate; he could trust that her offspring were also his and were therefore worth his effort to help raise. The basic outline of this scenario is similar to the current understanding, but with important differences. As was seen in chapter 2, humans' increasing brain size made nuclear families necessary because of the delayed development of young and the need for the male to invest in their support for many years. Trust in paternity was further inducement. Nuclear families emerged as an egalitarian partnership with mutual benefits to the parents and ultimate benefits to the survival of their young.

In 1966, Ardrey published a new book, *The Territorial Imperative*. He expanded on his ideas about baboons and included information about chimpanzees that had recently been reported by Jane Goodall, who had been observing them in Gombe, Tanzania, since 1960. He described baboons as "gangs of implacable thugs" whose aggressiveness among primates is second only to man's. He noted that the baboon male "is a born bully, a born criminal, and a born candidate for the hangman's noose." He is not only ugly, with "yellow-to-amber eyes that one associates with the riverboat gambler" (how scary is that!), but is also a thief that enjoys nothing more than killing. In brief, he wrote, the baboon is an "oversized monkey whose ways are uncomfortably reminiscent of man's" (p. 246).

Unlike the horrid baboon, Ardrey painted chimpanzees as models of amiability, following Goodall's lead at that time. He noted, "When band meets band…there is enormous excitement but no antagonism, and all may wind up feeding in the same trees." He opined that "the chimp is the only primate who has achieved that arcadian existence of primal innocence that we once believed was the paradise that man had somehow lost" (p. 222). One might have thought, given this wonderful endorsement, that Ardrey would consider a chimpanzee-type lifestyle for humans' forebears. This was not the case.

Ardrey concluded that species that succeed in evolutionary terms are those that have incorporated the qualities of "enmity, hostility, antagonism, [and] aggression" in their lifestyles. Baboons and human beings therefore flourish, but chimpanzees are on a downward path. They are the one social group "where amity seems the dominating force, drawing individuals together through nothing much but good nature" (p. 271).

In the mid-1960s, there was a pause in the baboon-chimpanzee controversy. During 1966 and 1967, three books came out that influenced a generation of zoologists and anthropologists. Two best sellers joined the aggro-men chorus that depicted human males and their predecessors as aggressive hunters, but neither was much interested in the details of how this came to be during evolution.

Konrad Lorenz, who was soon to win a Nobel Prize for his original research into animal behavior, in 1966 published the English version of his book *On Aggression*, which valorized aggression. He was not interested in human evolution per se and so had little to say about baboons or chimpanzees. One of his odd illustrations is about his old-fashioned aunt, who used to hire a maid to do household chores but would always fire her before the end of a year. The maid would be accused of various wrongs as the months passed and finally would be charged with some terrible crime, upon which the bewildered woman would be dismissed from her job without a reference. This may have seemed to Lorenz like a good example of a cyclic build-up of aggression in the aunt being periodically defused, but it could also be interpreted as the tale of a malicious old woman acting badly. Lorenz considered aggression to be an important part of humans' genetic make-up that erupts in everyone from time to time, but that can be assuaged by such things as athletic competitions or wars.

The second book was *The Naked Ape: A Zoologist's Study of the Human Animal* (1967), by Desmond Morris. Its folksy style made it a mass best seller. Some of Dagg's university students in the late 1960s decided to study biology because of it, and it was required reading in Harding's wildlife behavior class. Morris's focus was not on the hominid relationship with other primates but on why human beings had evolved to be the way they are.[6] He wrote that over millions of years, human ape progenitors had moved out of African forests to live on savannahs despite their inability to eat and digest the grasses and leaves that are consumed by herbivores. Their foods were no longer forest fruits and nuts but roots and bulbs, small animals such as mice and lizards, and scavenged meat. They already had good brains, good eyes, efficient grasping hands, and social organizations of some sort. About a million years ago, the males, having learned to walk and run on two legs, used their hands for wielding tools and weapons and planned together how to attack large animals; they were "a hunting ape, a killer ape…in the making" (p. 21). The females continued to be a drag on the males, as mentioned earlier. (They will have their revenge soon in this chapter, however.)

The third book, edited by Richard Lee and Irven DeVore (1968), comprises academic papers published from a 1965 conference on *Man the Hunter*; of the 67 attendees, only five were women. The papers describe the lifestyles of many indigenous cultures that lived by hunting, with little or no talk of baboons and chimpanzees, but the consensus of the meeting was that "the hunting way of life comprised 99 per cent of human history" (p. 341). Colin Turnbull made the interesting observation that it cannot be assumed that hunters such as the pygmies he studied are aggressive just because they kill animals. However, he also noted that husbands in pygmy camps often beat up their wives, just as their wives beat them up.

Sociologist Lionel Tiger carried on the theme of this conference in his 1969 book, *Men in Groups*, based on both a male "formalized hostility to females" and an attraction between males (p. xiv). Men certainly do form groups for such endeavors as fighting wars, playing contact sports, leading religious movements, and running large corporations and nations, but there is no proof that the tendency is derived from early hunting or defensive coalitions of protohominids rather than being drafted into an army, enjoying sports, or becoming immensely powerful and rich. Because of his theme, Tiger chose to valorize fierce baboons (which he discussed on 17 pages) rather than the less aggressive chimpanzees (which he mentioned on three pages). He also valued men over women, noting in the text that women have a pelvic structure that inhibits them "from the sort of locomotion useful and possibly essential for hunting" and observing that contemporary Bushmen stalking an animal may walk more than 20 miles in one day (p. 46). (Women who run marathons, take note.) Tiger further stated that women throw missiles such as spears with the same motion as protohominids (how could he know this?), a put-down that would surprise women playing professional baseball.[7]

Two years later in 1971, Tiger collaborated with Robin Fox to publish a book about humans' ancestors and human nature entitled *The Imperial Animal*. As academics, they must have been aware that human beings are much more closely related to chimpanzees than to baboons, but one would not know this from reading their book. They paid far more attention to the monkey than the ape, devoting 6 times as many pages to the baboon, which they called "our cousin." After all, baboons are highly aggressive. Barbara Smuts (1985), who later observed olive baboons in Kenya, reported that each female in her troop was threatened or chased by males on average once every 17 hours. In about a quarter of these episodes, the female was actually assaulted. Such attacks, about one a week for each female, were painful. A leg wound might make the female limp for days. Some wounds were 10 to 15 centimeters long, and one young female died from a gash on her abdomen that exposed her entrails. Tiger and Fox's reasoning for valorizing baboons seems obviously to be because they are vicious (shown by Zuckerman and field studies), whereas chimpanzees are peaceable (reported by Goodall in 1965). By using the baboon as their example of possible human ancestors, they could tie in aggression with human nature, the theme of their book.

Needless to say, a model of male aggression that implied fights and dominance hierarchies in primate groups was pervasive at this time. When Shirley Strum (1987, p. 161) organized a conference in 1978 at which she announced that the male olive baboons she had observed for years had no such hierarchy (although the females did), there was disbelief among the other attendees. (Thelma Rowell had reached this same conclusion in 1966 but had been too reticent to push her findings against the powerful male triumvirate of Zuckerman, Washburn, and DeVore [Jahme, 2001].) Almost all the conference attendees fed into the paradigm that male dominance hierarchies within their societies were essential to keep male aggression in check. They insisted that Strum had invented her results, or had not understood what she was seeing, or had not collected enough information. Then a man gave a paper based on far less research than hers about three different troops, all of which he said had a male dominance hierarchy. Even though his study was not nearly as thorough as hers, his work was accepted with accolades by the primatologists who were present. They knew there *had* to be such a hierarchy, even though future work would prove them to be wrong and Strum and Rowell right. The women's research would help create a new paradigm shift.

GENDER WARS

Academic women had put up with enough nonsense about the preeminence of males in hominid prehistory. It was time for a revolution. Louis Leakey had known what he was doing when he recruited three women, Jane Goodall, Dian Fossey, and Biruté Galdikas, to research the behavior of three great apes, chimpanzees, gorillas, and orangutans. None of these women had an academic background in zoology or animal behavior, so all came to the research with open minds, ready to describe the behavior they observed rather than try to make what they saw fit the current hypotheses.

During the 1960s and 1970s, women academics had become more common and increasingly active in working out their own theories of human evolution. They devised scenarios that seemed more logical than those proposed by men. Nancy Makepeace Tanner in her book *On Becoming Human* (1981) discussed humans' early hominid ancestor *Australopithecus* evolving some 8 million to 4 million years ago on the savannah lands of Ethiopia, Kenya, and Tanzania. *Australopithecus* were far too small to be real hunters; they were different from forest apes not only because of their open habitat but because they walked on two legs and used their hands and tools to scavenge and gather food over a large range. The females were the more innovative and involved sex in the mating game because they had their children to attend to as well as their own nutrition so they could produce and nurse healthy babies. The most effective females, meaning those who had a higher number of healthy progeny that survived to maturity to reproduce in turn, were presumably the most intelligent and energetic. Such females would choose to mate with progressive males and decline to mate with others. Gradually, a system of egalitarian couples within each nomadic group evolved, along with advanced bipedalism, hand manipulation ability, sociability, and intelligence. Because the male knew that the children his partner produced were his own, he was willing to help raise them. Male forebears had had large canines for fighting, but females were not interested in aggression and would choose mates with small omnivorous teeth like their own. Indeed, during further evolution, sociable male and female hominids began to resemble each other more than had their ancestors. Today (and for the last 4.4 million years, back to *Ardipithecus*), men and women have similar teeth and other anatomical systems, although there remains a small discrepancy (15–20%) in size and weight.

The feminist–aggro-men debates had begun. Many academic men were not pleased with a hypothesis such as Tanner's that valorized women's work and a peaceful life during human evolution. Richard Wrangham and Dale Peterson in their book *Demonic Males: Apes and the Origins of Human Violence* (1996) reiterated the arguments already made about the primacy of male aggression and male "mighty hunters" in human prehistory.[8] They portrayed men not as beings who might organize society in such a way as to bring peace and prosperity to its members but as hardwired demons who, with their high brain power, can think of ever new ways to manipulate "new weapons, new tactics, new kinds of deception in the ever-escalating game of conflict" (p. 257). They also concluded, to the annoyance of feminists, that "we should accept the likelihood that male violence and male dominance over women have long been a part of our history" (p. 251). Wrangham and Peterson did acknowledge, despite the thesis of their book, that it is "true that for the most part chimpanzees lead very peaceful lives" (p. 11).

As if to back up the new book on demonic males, also in 1996 the documentary film *Warriors of the Amazon*, about the Yanomami tribe of Venezuela, was premiered as a Nova/BBC special. Patrick Tierney (2000, p. 217) in his book *Darkness in El Dorado* noted that Nova was determined to portray this tribe as a warrior society to titillate audiences, which it did. The narrator stated that men went off to fight two or three times a year, although this was not in fact true; indeed, one of the few "wars" at the time was between villages dominated by two anthropologists, Jacques Lizot (an active pedophile) from France and American Napoleon Chagnon, both men who had a negative effect on their subjects of study. (These men also ignored, devalued, and/or stole information from the only white person who had lived for many years with the Yanomami and who spoke their language, Helena Valero.) The film showed a mother overseeing her sons hurting each other to encourage violence and Yanomami boys burning themselves for conditioning against pain. American missionaries watching the film who had worked with tribe members for a lifetime laughed at this, claiming that Yanomami were the "biggest babies in the world," (Tierney 2000, p. 217), making a huge fuss if they suffered a small bruise or cut. The film crew gave cooking pots and machetes to the locals for their participation in the film, but they chose not to help a woman and her newborn baby who were filmed week by week as they slowly weakened, died, and were cremated. Yanomami people were angry when they saw the film that was supposed to portray their lives. Tierney called it "a misleading film about a fake feast that co-generated a dangerous new military alliance" (p. 220). With this film, misleading facts about human aggression were reaching a wider, nonreading audience.

In 1999, Craig Stanford, another aggro-man, published *The Hunting Apes*, which gives a new twist to the same theme. He claimed that the sharing of meat obtained by hunting was the pivotal agency in human evolution (although meat for chimpanzees today is actually a luxury, accounting for about 2% of an individual's weekly diet [Cohen, 2010]).[9] Stanford claimed that mostly male coalitions hunted meat that they used to exchange favors for mating (a fallacy that we discuss in chapter 6), thereby setting up a relationship among males and between males and females. His example is of necessity the chimpanzee, because the other apes (except occasionally bonobos and gibbons) are not interested in hunting or meat. Most chimpanzee females like meat, but they certainly do not need an incentive to mate. They are sexually liberated, as demonstrated in Jane Goodall's documentation of one female in estrus cheerfully mating with 14 males on one day. Stanford's book is not persuasive. About the same time (in 1998), he published a paper arguing that chimpanzees and bonobos are not really that different after all (the topic of the next chapter), perhaps so that fewer critics would object to his choosing chimpanzees rather than bonobos as most like humans' ancestors. However, a variety of primatologists disagreed emphatically in comments appended to his paper.

In 2005, Donna Hart and Robert Sussman published *Man the Hunted: Primates, Predators, and Human Evolution*, the most logical account of human evolution yet put forth. They focused, as did Nancy Tanner, on the 3 1/2-foot-tall female australopithecine named Lucy who lived in northeast Africa 3.2 million years ago. The photo of her small skull and little bones spread out in skeletal order on a table does not resonate at all with the title "hunter," unless roots, clams, and fruit were her prey. Her male peers were little larger than she, both with small canine teeth that could not have been used for fighting. Rather than hunting, these small beings themselves were prey for the ancestors of lions, tigers, leopards, wild dogs, hyenas, snakes, crocodiles, sharks, and even eagles—animals whose present-day descendants still kill many people in Africa each year, as black bears, grizzly bears, and mountain lions rarely do in North America. The authors reported that stone tools but not weapons had been found with fossil bones of hominids from 2.3 million years ago, and that evidence of large-scale human hunting appeared only about 60,000 to 80,000 years ago, after 99% of human evolution was completed.

Ignoring this information entirely, in 2007 David Livingstone Smith published his book *The Most Dangerous Animal: Human Nature and the Origins of War*, which reverts again to the aggressiveness of humans' ancestors. He wrote, "We are social primates, and share the xenophobic and violent tendencies of our chimpanzee cousins" and that fierce males "transmitted a proclivity for male coalitionary violence down the generations until it saturated most of humanity" (p. 92). Smith agreed that chimpanzees and bonobos are equally human beings' closest DNA relatives, but he chose chimpanzees as being more like people because, he asserted, humans and their ancestors, like chimpanzees, are and were aggressive and violent (the old story), and also because "chimpanzees are more like gorillas than bonobos are" (whatever that means) (p. 79). As was seen in chapter 2, there is no evidence in the archaeological record for human violence against humans until about 20,000 years ago and no evidence for mass killing until 12,000 years ago. To say otherwise is pure fantasy.

WHAT ABOUT HUNTING?

Aggro-men emphasize hunting for humans' early ancestors both to distinguish them from other primates and to valorize their aggression. Most primates such as monkeys, gibbons, orangutans, and gorillas do not include red meat in their diet. Chimpanzees, bonobos, and gibbons sometimes eat small animals they can catch, but their consumption of meat is minimal. Scavenging meat for them is almost unknown. Outside the developed world, most people do not eat much meat, either.

That human beings evolved basically as plant eaters is evident from their biology (Hart & Sussman, 2009):

1. Humans' teeth are not designed to eat meat. Lions have large canine teeth and fixed lower jaws that give their bite strength as they shear off meat, whereas humans' lower jaws move easily back and forth and side to side, enabling them to bite and grind plant material. A lion can devour a large chunk of meat in a few seconds, but humans must chew food patiently—preferably 32 times a mouthful, according to the early health advocate Émile Coué.

2. The intestinal tract of carnivores is short and straight, whereas that of people and other plant eaters must be long and coiled so that the nutrients from plant food can be absorbed as they move slowly through the gut.

3. As humans chew, the salivary enzyme ptyalin combines with complex carbohydrates in the mouth to start breaking them down to simple sugars. Ptyalin is not present in carnivore saliva.

4. A highly concentrated acid in the stomachs of carnivores breaks up the muscle and bone they have "wolfed" down, whereas the stomachs of humans and other plant eaters contain much lower levels of acid.

5. Cholesterol is manufactured in the liver of humans and carnivores and is also consumed in food derived from animals, such as meat, fat, eggs, and milk. Carnivores readily process and excrete high amounts of cholesterol in their feces, but the livers of humans and other plant eaters are highly inefficient in removing dietary cholesterol. Instead, it enters humans' bloodstream and can be deposited in the arteries, triggering many diseases, such as coronary heart attack, stroke, diabetes, and cancers.

Before the evolution of bipedal terrestrial hominids, humans' early ancestors were mainly fruit eaters, as gibbons (Elder, 2009), orangutans (Galdikas, 1988), bonobos, and chimpanzees are today. This feature distinguishes apes from Old World monkeys, which have a digestive system design, dentition, and other features that allow them to consume a much higher proportion of fibrous leaves, unripe fruit, and tough nuts (Norton, Rhine, & Wynn, 1987; Scott, 2011). After the split, gorilla ancestors evolved a more leaf-based diet (Rogers, Abernethy, Bermejo, Cipolletta, Doran, Mcfarland, Nishihara, et al., 2004).

More recently (given the huge time scale of evolution), culture has had a large effect on the protohuman diet (Hart & Sussman, 2009). Cooking allows individuals to chew and digest raw muscle fibers from meat and originally inedible vegetables, so when protohumans developed the ability to control fire, they could use it to greatly expand their diet. This was accomplished by the large-brained *Homo erectus* about 790,000 years ago. (Earlier dates have been suggested but cannot be verified.) Fossil and archaeological evidence indicates that *Homo erectus* developed large-scale, systematic hunting of large animals only between 60,000 and 80,000 years ago.

WHY THE BIAS OF AGGRO-MEN?

Because aggro-men by definition buy into human aggression, they valorize the fierceness of chimpanzees, their choice for the ape that is most like humans. This is amazing because many researchers have reported that this ferocity is in large part (it cannot be known how large) a result of human intrusion into chimpanzees' lives—that is, through the offering of free food that weaned chimpanzees away from their natural diet but then was produced with a randomness that frustrated those that became accustomed to this largess. (People of third world countries being fed by the United Nations after suffering from some horrendous disaster express similar rage when shipments of food do not arrive or are too small. Who can blame them?) Chimpanzee researchers Richard Wrangham (1974) and Vernon Reynolds (1975) were both personally aware of this problem, as was their colleague Jane Goodall (1986). Primatologists Margaret Power (1991) and David Hill (1994) have discussed it at length (see chapter 4). Yet the behavioral changes that are induced by provisioning are quickly dismissed or ignored entirely by aggro-men.

Because aggro-men love hunters, too, they are biased toward the idea of protohominids being hunters rather than forager-gatherers. Hart and Sussman (2005) argued that most academics come from a Judeo-Christian background that valorizes the dominance of man over nature. Most scholars have been men, and they chose to write books and articles about the prehistoric prominence of their gender, as represented by their hunting ability. They resisted the thought of women being important in society because they gathered edible vegetation, although it has been known for four decades that some 60% to 90% of the foods that are used and collected by most modern-day human foragers in the tropics is still provided by women. The earlier significance of Piltdown Man (proven a fraud only in 1953) had also predisposed scientists to think of humans' early ancestors as having a large brain that would have enabled them to become skilled hunters long before that actually happened.

Where meat is concerned, modern humans' early hominid ancestors can be thought of not as hunters but as scavengers who got protein from small animals such as insects, lizards, and birds' eggs that they found while foraging for fruit, as frugivorous gibbons do. Louis Leakey recounted how he and his son, Richard, reenacted the possibility of scavenging on the plains of Africa. They took off their clothes, grabbed giraffe leg bones to protect themselves, and approached hyenas feasting on a zebra kill that lions had made. They were able to scare away and hold at bay the hyenas for about 10 minutes while they hacked meat from the zebra. The hyenas were furious, Leakey and Ardrey (1971) reported. Such meat was probably not an important source of food in the early evolution of humans' ancestors, though, given that in warm weather, it quickly becomes putrid with bacteria, maggots, parasites, viruses, and fungi (Hart & Sussman, 2009).

Until recently, humans' male ancestors have been thought of as aggressive. Fossil bones have been found with arrowheads embedded in them, massed graves of bodies slain by victorious males have been uncovered, and history books tell of ancient and recent massacres and battles. But such evidence is from thousands, not millions of years ago, after humans' ancestors had developed a large brain that enabled them to figure out that "might is right" (practically but not ethically) and to make weapons and organize the slaughter of enemies. Before that time, for millions of years, male protohominids would not have been any more aggressive than other apes and were probably less so given their small size and predator-laden terrestrial habitat. When males evolved as partners to women, they became not aggressive hunters but caregivers to their children: family men. And they evolved in this role for millions of years.

Endnotes

1. How these men classified the races of humans is another aspect of bias, but one that is thankfully beyond the scope of this book.
2. Zuckerman planned to call this book *The Sexual Life of Primates* until he was reminded that "Primates" (with a capital *P*) referred to archbishops or bishops of the Anglican Church.
3. This experiment is being repeated today on a massive scale in China, where abuses of the one-child policy have resulted in some 33 million excess men who cannot find mates and most of whom cannot leave the country (Hesketh, 2009).
4. In keeping with the theory of male dominance and of males as "fighting machines," Irven DeVore reported that baboon troops traveled from place to place in an "army model," with nursing females and infants in the core, surrounded and protected by high-ranking males and females. Adolescent males moved on the periphery of this group, deflecting attacks and keeping the most vulnerable members safe (Jahme, 2001, p. 93). It is now known that this macho arrangement was a complete fantasy; when a baboon troop is threatened, all the animals take flight to save their own skins. The larger males run faster than the females carrying infants, so it is the latter who are likely to be taken by leopards.
5. Many other popular books have been written about human evolution, such as Jared Diamond's *The Third Chimpanzee* (1992) and John Gribbin and Jeremy Cherfas's *The First Chimpanzee* (2001), but these focus primarily on such things as bones, tools, genes, and indigenous cultures rather than on ephemeral behavior as it existed for millions of years.
6. Morris did come up with wild and amusing conjectures, though, among them the notion that our predecessors may have lost their fur because eating meat was messy; their body hair would have become clogged with blood and gore and therefore been a disease risk. For the same reason, vultures lost the feathers on their heads during their evolution. *The Naked Ape* became a mass best seller in part because of its ideas about religion. Most notably, Morris described religion in a way that shocked readers although it was based on the idea of pack hunters needing dominant leaders to organize forays and subordinates to carry them out. Morris claimed that religion had such amazing potency not because it came from God but because it reflected the fundamental hierarchical system that is

present in most monkey societies. Human beings, like their primate ancestors, had inherited the biological imperative to submit to a higher power, but in humans' case, this submission was not to the alpha male ape but to a male god, as articulated by influential men who knew a good opportunity when they saw it. Morris argued that these men dreamed up a number of "bizarre by-products," among them the belief that there was another life after death where people could meet this God as well as family members who had died before them. Morris claimed that this ideal of submission was so ingrained in human beings that if one religion were destroyed, another would spring up in its place. Thus, the idea that "man" evolved as a hunter was piggybacked on that of religious blasphemy. Morris believed that aggression, a particularly male attribute, surfaces whenever people are tense or stressed, so that many small rituals have evolved to keep it in check. Say, for example, a man goes to a fancy soiree. His agitated state of conflict about this brings out an assortment of displacement activities. He stubs out a cigarette he has just lit. He polishes his glasses although they are already clean. He sips a displacement drink although he is not thirsty. He nibbles displacement food without being hungry. He glances at his watch but does not register the time. All of these actions appease the "hidden fears and aggression lurking just below the surface" (p. 148). In people's most intense moments of aggressive tension, Morris claims that they revert to activities similar to those used by their ape relatives. Chimpanzees scratch in a stylized way, just as people scratch, bite their nails, smooth their hair, rub their chin, or lick their lips. One would have thought these actions were the result of nervousness, but Morris claimed instead they are necessary to circumvent the aggression that might otherwise erupt.

7. A large body of literature, especially since the 1990s, compares male and female athletes' physiology, physical structures, performance, and injuries. Key points are that men have greater muscle mass and longer bones that give greater strength and leverage in the upper body, but these differences essentially disappear at the waist. There are minor differences in hip structure that load stress on the bones, ligaments, and muscles of men and women a little differently, but for runners, individual differences in cardiovascular and muscle efficiency (both inherent and those resulting from training) are far greater than the average gender differences. A glance at Olympic and World Record statistics shows that in events from the 100 m sprint to the marathon, men run about 8% to 11% faster than women. But, since human males are 15% to 20% bigger (depending on whether

one measures weight or height), this means that, pound for pound, elite female athletes are 5% to 10% faster than males, based on the principle of allometric scaling.

8. Wrangham was present as a young researcher at Gombe when the male chimpanzees of one community violently attacked and killed members of another, which must have been a searing experience for him.
9. Jon Cohen (2010) argued that diet was fundamental in making human beings unique, but not through the eating of meat, because relatively little meat was and is available to chimpanzees. Rather, human longevity, which is much greater than that of other apes, is correlated with the grandmother hypothesis of Kristen Hawkes (2003); this postulates that food collection and child care by postmenopausal females helps their daughters to produce more babies and to raise and teach them over a longer time period. It also correlates with Richard Wrangham's cooking hypothesis (2009), whereby the control of fire enabled humans' ancestors to access more kinds of food, make this food more digestible and tasty, and drastically reduce the time that was needed each day to find food.

Chapter 4

Humans' Closest Cousins —Not That Fierce

Even though, as was seen in chapter 2, human beings did not evolve from a chimpanzee-like ancestor and have a vastly different mating system, aggro-men continue to use chimpanzees as a model for human behavior. Therefore, we shall consider them in detail in this chapter, along with their truly closest relatives, the bonobos.

Although Jane Goodall was all too aware of the "war" that had convulsed her chimpanzee kingdom, she still was able to write later (in her 1986 compendium of chimpanzee behavior) that peaceful interactions between individuals "are far more frequent than aggressive ones; mild threatening gestures are more common than vigorous ones; threats *per se* occur much more often than fights; and serious, wounding fights are very rare compared to brief relatively mild ones" (p. 357). As far as quarrelsome primates go, chimpanzees, especially those that have been randomly provisioned by researchers, are high on the continuum. But that does not mean that chimp life in general does not flow along peacefully. It does, just interrupted now and then by antsy and excitable males. There are squabbles and spats, but few individuals are wounded and killing is almost unknown within a group except, rarely, for infants. Most other primates are even less excitable than chimpanzees.

Male chimpanzees are considered a poster species for aggression. Their ferocity has been valorized by many authors and connected, without evidence, to aggression in humans' early male ancestors. Because chimpanzees are aggressive, these aggro-men writers aver, early hominid males millions of years ago must have been aggressive, too. This book reconsiders this odd belief. Why should men be considered fundamentally aggressive because male apes living today in Africa are thought to be? Even the assumption—the myth, more accurately—that chimpanzee males are very aggressive fails when data on their behavior are studied in detail, as Margaret Power did in her thoughtful book *The Egalitarians—Human and Chimpanzee* (1991).

Power spent many years mulling over early versus later data collected in Africa on chimpanzee behavior. In the early 1960s, two main research sites were being developed, both in Tanzania: Gombe, founded by Jane Goodall, and the nearby Mahale Mountains, where Japanese scientists worked (Nishida, 1990). At both sites, chimpanzees were soon given free food so that they would become used to the human voyeurs who provided it, which is what happened. Although Louis Leakey was furious when he found out that his protégé Goodall was feeding the chimpanzees, she felt she had no alternative because of funding: The *National Geographic* was paying her for excellent photos of the chimpanzees, for which she needed bananas to bring them close (Peterson, 2006). However, as the years progressed, this free food program would have huge repercussions that would convulse chimpanzee communities—not because of the food itself but as a result of its methods of presentation, which caused widespread frustration and anger. Power hoped that further field studies where the apes were *not* given food would help prove her hypothesis, and indeed, they have in such research as that of chimpanzees in the Budongo Forest of Uganda, which have never been provisioned (Reynolds, 2005).[1]

Margaret Power described the various successive plans Goodall and her team used in provisioning the chimpanzees at Gombe. Plan A involved giving bananas to the few chimpanzees that first ventured into her camp. Slowly, the number of chimpanzees increased, so more fruit had to be imported and set out—an expensive treat because one adult could eat 60 or more bananas at a sitting. Baboons came to the feast, too, freeloading food with abandon.

To combat the unwanted baboons, in 1965 Goodall tried Plan B. This involved embedding concrete feeding boxes into the ground. Using a series of buried wires, a researcher some distance away could open a box of fruit when this seemed desirable, such as when there were few baboons present. However, whenever a box was opened, all the chimpanzees converged on it at once, fighting to grab as much fruit as possible. Over the years, the large males that grabbed most of the food became increasingly aggressive.

There were other changes, too. Chimpanzees came in larger groups, they often slept near the feeding site in order to get to the banquet as soon as possible, and they frequently hung around the camp for hours (Van Lawick-Goodall, 1971; Power, 1991). Whereas earlier, their fission-fusion type of community had meant that individuals came or went as they pleased along with a variety of short-term comrades, now the larger, more stable groups were dominated by the big males. At this time, chimpanzee David Greybeard would take a researcher by the hand and lead him or her to a closed box of bananas. He did not realize that the box could be opened only from a distance, one researcher reported: "You sit with him at the box held in his iron grip, and he gets progressively more livid because you don't open the box" (Peterson, 2006, p. 392). That people and the chimps were actually touching each other was upsetting to Goodall because they could pass infectious germs back and forth between them, but it was difficult if not impossible to prevent this.

By 1967, there were 58 chimpanzees coming fairly regularly to the feeding site despite the pervasive tension and aggression. Frustration ramped up when the animals could smell the food and knew it was in the boxes but could not get at it. At this time, the concept of a database for chimpanzee behavior was inaugurated, which enabled researchers to quantify behavior. They could no longer report that chimpanzees were friendly and shared food, because this was no longer the reality. The database on the natural behavior of chimpanzees at the food site would be badly biased from its very inception because of human interference.

In August 1967, Goodall introduced Plan C to try to stop the mayhem. The boxes now provided bananas for a day or two before being left empty for a few more days in an unpredictable sequence. This random schedule was supposed to reduce the time that the chimpanzees hung around the camp, but in fact, it increased aggression on the days that bananas *were* available, not only among the chimpanzees but between them and local baboons. When Goodall returned to Gombe after some time away at university, she found the feeding situation in chaos. Leaving the boxes closed "merely built up tremendous tensions and frustrations in the chimpanzees who were there," she wrote (as cited in Power, 1991, p. 29). The longer the boxes remained closed, the worse the situation became. Finally opening the boxes at last created "bedlam" (cited in Power, 1991, p. 30) among the animals.

In 1968, Plan D came into effect, which was no better. In it, each individual was to be fed about one box of bananas every 10 days or so at the discretion of a researcher. But again, the apes could smell the bananas as they waited about, prying at the boxes with sticks, and again they were incredibly frustrated. Wild chimpanzees would have spent the days wandering companionably through the forest from one food source of leaves or fruit to another. By contrast, these provisioned animals sometimes wasted days in the camp during which they received no food at all. To have to go hungry was an unnatural and horrific experience for them.

From 1969 to at least 1972, Goodall's team tried Plan E. This involved a researcher, often a student, giving bananas out in theory once a week to each individual. However, the animals arrived in groups, all hungry, and this objective was impossible to achieve. Provisioning was by now obviously a terrible idea if one wanted to study the natural behavior of a wild species. It had turned small groups of affable apes into larger groups of aggressive and hierarchical individuals. It promoted bullying and brutality.

Power emphasized the point that of the hundreds of people who have visited and observed the chimpanzees at Gombe, only Goodall and several others had known the animals as they were before 1965, under wild conditions in which they shared food and roamed in small, friendly groups. All the other observers thought of the males as aggressive, which they had become.

Power also described the provisioning routine at the Mahale study site in detail; it too turned animals that were once friendly and peaceful into aggressive and frustrated individuals. The researchers tried growing sugarcane for the chimpanzees, but the ripe crop was ruined by their overfeeding. They then provided cut sugarcane, but this also elicited major aggression and frustration when there was not enough food to go around. Sugarcane was sometimes frugally distributed, then withheld for an hour while the researchers recorded the behavior of feeding individuals. This method was particularly irritating to the chimpanzees because they knew that people, not boxes, had the food, but refused to deliver it. The apes might hang about for hours, yet still receive little to eat. Their frustration increased.

At these two sites where providing free food had been extensive and chimpanzees were thwarted during weeks and months and years of expecting free food and often not receiving it, meaning that they went hungry, frustration was rampant. The large males became aggressive, grabbing most of the available food, roughly pushing infants aside and demanding instead of requesting sex from females. Instead of playing with young baboons as in the past, chimpanzees sometimes killed them. At Gombe, one provisioned community gradually split into two with separate territories. It was one of these that, its social cohesion destroyed by long-term frustration and unnatural grouping, eventually killed off members of the other. The same scenario seems to have been played out at the Mahale site. Because of frustration, the behavior of chimpanzee communities changed quite quickly from amiability to aggressiveness. Most social animals have the capacity to be aggressive, but this behavior is detrimental to group life, and under normal conditions it is little in evidence.

Besides frustration and the associated pathological behaviors, this history may reflect learned behavior. The big males were rewarded for being aggressive and became more so. The smaller individuals learned either to keep out of their way or to assist them, reinforcing the big ones' antisocial behavior and promoting unnatural coalitions.

There are 40 years of data each from Gombe and Mahale post-1965 and post-1968 observations illustrating how aggressively provisioned chimpanzees may behave. Fortunately, Goodall (1965) also wrote up a chimpanzee report from her first 24 months of observations between 1960 and 1962, when free food was seldom given to the chimps. The difference between the two sets of data is great. During the early years, Goodall observed one animal acting as if it were dominant to another 72 times, with three quarters of these interactions taking place between two males. About once every 10 days or so, for example, two chimps might meet on a branch and one would give way. There was one fight between adult males, but they were generally very tolerant, grooming each other much more often than other combinations of chimps did and occasionally even sharing food. Goodall recorded no dominance hierarchy and no permanent leaders. She did, however, report four cases of threatening behavior, shades of future dysfunction: Three of them were triggered when a banana was thrown between two males. The larger of the two marched toward the other giving "high-pitched screams, his mouth half-open, lips drawn back from his teeth, and the hair of his shoulders, back, and arms erect" (p. 466). Bananas were obviously taken seriously by the apes—as they should have been by the researchers, had they known what would happen. When Goodall published her book *In the Shadow of Man* (Van Lawick-Goodall,

1971), she was already reporting the chimpanzees' excessive aggression compared to the early days: In the index, "aggression" is recorded on 68 pages. What a pity that in her later reports, Goodall assumed that the early friendly data she had documented had been misleading and decided that the later, more aggressive interactions reflected true chimpanzee behavior and therefore were what she included in her 1986 definitive 670-page book.

Goodall was horrified to have to announce in 1974 that after 14 years of watching her chimpanzees live a more or less peaceful existence, some males had become murderers. The large group of chimpanzees she had observed had split into two, which then inhabited separate territories. The more powerful group had turned against their erstwhile friends and killed them in cold blood over a period of time. Goodall referred to this catastrophe as a "war." This concept of war has been repeated many times by aggro-men to strengthen the position of the camp that believes men (like chimpanzees) are inherently aggressive.

What is to be made of chimpanzees being at "war"? From a rational perspective, Goodall's chimpanzees had not been involved in a war, which is a human activity defined as an armed conflict between tribes or nations. Goodall had seen a group of males killing members of another group—a dispute of a kind that is present in various species, usually involving territory and resources. Goodall herself was undoubtedly appalled to realize that the "war" (1986, p. 506) was surely caused in part by the provisioning of bananas she herself had organized (Reynolds, 1975; Wrangham, 1974).

It is known how truly wild (nonprovisioned) chimpanzees behave from data collected in the early (preprovisioned) years at Gombe and Mahale and from other research sites such as that in the Budongo Forest of Uganda at Sonso. There, where the apes were never given free food, these animals live a far more peaceful life from year to year than do the Gombe animals. Vernon Reynolds, a distinguished professor at Oxford University, set up a research station at Sonso in 1990. During the next decade, he organized research help from 138 other workers, most of them graduate students or senior scientists (Reynolds, 2005). In his definitive book *The Chimpanzees of the Budongo Forest*, Reynolds reported that the males at Sonso "can be tremendously friendly and mutual with one another, and move around together in seemingly perfect harmony" (p. 124). In 1 year of collecting behavioral data, there were only 28 instances of aggression, which comprised merely threatening and chasing. Males groom one another far more than do females, actively seeking out preferred friends for this activity. Mutual grooming is an ancient, ubiquitous behavior in primates that universally promotes interpersonal harmony and group cohesion (Dunbar, 1991). Male friends also serve as generalized fathers, which makes sense—probably all have mated with the mothers of infants and could well be their father (male-male interactions are further explored in chapter 9). Their tolerance of these youngsters is amazing; adult males allow infants to crawl over them when they are resting, sit beside them when they feed, and even share their food. Males even put up with infants trying to interfere with their copulations. Reynolds noted that the males infrequently threaten or chase females, but these squabbles are easily contained. However, rarely real aggression can occur: On one occasion, a presumptuous young adult male, Zesta, was killed by his mates, and there were several cases of infanticide.

As Goodall, Reynolds, and Power have reported, a truly wild, unprovisioned chimpanzee community looks like this: Each has a loose home range that overlaps that of other groups. Sometimes two groups meet in a carnival-like rush of exuberant cacophonous excitement where friendships are renewed and couples may mate. In their fission-fusion lifestyle, individuals mainly travel and forage in small subgroups whose composition changes frequently; if two chimpanzees want time away from each other, they can simply join different subgroups to forestall possible tension or aggression. Subgroups are of two kinds: those that travel slowly because they include infants and those that are more mobile, cover more ground, and include both males and the few females unencumbered by offspring. When a mobile subgroup comes upon a rich source of food such as a large fruiting tree or a grove, as both Goodall and Reynolds reported, the members do not think of keeping it for themselves, but rather make a terrific racket hooting and drumming on tree trunks so that nearby chimpanzees can come and join in the bonanza. All individuals come and go as they please. Researchers reporting on these wild chimpanzees have used emotive words to describe them, such as "frictionless peace," "happy-go-lucky animals," "carefree," and "extroverted, unconcealed joy" (Power, 1991, p. 57).

Power compared the social organization of unprovisioned chimpanzees with that of the nomadic !Kung (also called San) tribe of Botswana as it was in the 1950s, before the members settled down. This was probably how human ancestors existed for millions of years until the beginning of agriculture about 10,000 years ago. Like chimpanzees, the !Kung foraged for food and lived in loose home ranges that overlapped those of other small groups. They seldom fought with each other, had no laws, and ruled themselves by a reward-and-punishment system. Group members who behaved in a socially positive way remained in friendship with the rest of the community. If a person behaved badly, he or she would be sanctioned with shaming, mocking, or shunning. Adults quickly intervened when children's play threatened to get out of control so that they never experienced physical violence (Draper, 1978). This makes sense, given that aggressiveness within a group is counterproductive to its smooth functioning.

By contrast, other indigenous peoples fostered aggression by training boys to fight. Mae Enga tribesmen in the New Guinea Highlands encouraged boys from the age of eight to move to their fathers' huts rather than stay with the women. There they were allowed to scuffle and brawl, and later to have mock battles against each other in teams to learn military tactics that would be used later against other tribes (Meggitt, 1977). By the time they were adults, they were ready to join the group's men as warriors, able and willing to fight when the occasion demanded. The men of Papua New Guinea remain violent to this day, often taking out their aggression on their wives; two thirds of rural women and more than half of urban women have been beaten by their partners (Toft, 1986). A common injury for them is having their forearm bones snapped because they tried to protect their heads from a blow by a club or other weapon (Dubois, 2010). Similar violence also occurs today in the developed world, where many men are overwhelmed by poverty and a lack of jobs; frustration for them is as devastating as was the random provisioning of chimpanzee communities. In urban slums, drug dealers, pimps, prostitutes, fistfights, and gun battles flourish. Aggressive behavior in both humans and animals is a "complex interaction between genes and environment, with social experience playing a crucial role" (Montagu, 1978, p. 5). Aggression in human societies is obviously highly variable.

If one were to choose a species of monkey or ape living today that would probably have been most like modern humans' progenitors, which would it be? It could not be any monkey, because their skeletons, physiology, and behavior are quite unlike humans'. As for the apes, it could not be the orangutan, gibbon, or gorilla, because fossil evidence shows that their ancestors split off from humans' hominid line millions of years earlier than did those of the chimpanzee-bonobo line (although we showed in chapter 2 that the human line evolved from a form something like a gibbon). Gibbons are a tempting model, however, because, like humans, they are not sexually dimorphic, have concealed ovulation,[2] live in nuclear families consisting of a mated pair and their dependent offspring of several ages, and mate for life, although with a low prevalence of infidelity (or as biologists say, "extrapair copulations"). But gibbons also exhibit some differences from humans; for example, the pair defends a territory that they almost never leave against other gibbons, and they copulate infrequently, only when the females come into estrus.

Notwithstanding humans' close relationship to chimpanzees and bonobos (in the recent evolutionary divergence and about 98.9% similarity in coding DNA[3]), they also are a poor model for human behavior, or at least for mating behavior and intragroup aggression. Chimpanzees and bonobos both have promiscuous, albeit different, mating systems and commensurate physiology and behavior.

All recent authors writing about human evolution would agree on these facts. But interpretation is everything. No matter what a species' attributes, for aggro-men aggression seems to trump all.

With Goodall's mass media bombshell in 1974 that some male chimpanzees went to "war" just like people, it became evident to these aggromen that the aggressive chimpanzees could now be considered as most like humans' early ancestors. That chimps were apes with DNA very like human beings' own was especially important to them. The fervor was so strong that several primate taxonomists proposed reclassifying them in the *Homo* genus with human beings (Diamond, 1992; Goodman et al., 1998; Watson et al., 2001); this arrangement was quickly discarded—an indication of the ebb and flow of scientific opinion.

But what about the other species of chimpanzee, the pygmy chimpanzee now called bonobo? They are as closely related to humans as are chimpanzees, and research shows that they are less aggressive than chimpanzees, settling most of their problems with sex rather than swats. Bonobos have been seriously studied since 1973, with exciting books about them in English (rather than Japanese because Japanese men were carrying out the basic research) available by the 1990s (de Waal, 1997; Kano, 1992). Surely bonobos were at least as likely as chimpanzees to behave as did humans' ancestors? Yet men continued to write books about aggressive man and Man the Hunter based on humans' early ancestors, who, they fantasized, were aggressive like chimpanzees (Smith, 2007; Stanford, 1999; Wrangham & Peterson, 1996), as noted in chapter 3.

Humans' Closest Relatives: Chimpanzees or Bonobos?

Are human beings more closely related to chimpanzees or to bonobos? These apes have DNA almost identical to each other and very similar to humans' own. Their common ancestor is thought to have lived between 2.7 and 1.2 million years ago (Boesch, 2002, p. 4). Both inhabit tropical forests in Africa, but their environments are different. Chimpanzees live where food is hard to come by, so rather than foraging in a group, mothers with their young usually wander and feed on their own wherever they can find edible fruit, flowers, or leaves. Because of this, females are not closely bonded. By contrast, bonobos in Congo have abundant food sources from which to choose. Females forage together with their young, forming strong friendships. These differences have a huge impact on their behavior. Chimpanzee females are smaller than males, so males are dominant and can boss them about. Female bonobos are also smaller than males, but this does not matter. If a male tries to pull rank on a female, her sisters rally immediately to her aid. Female bonobos are in effect dominant to males, because they band together to get what they want.

We will consider what these two species have in common and then how they are different, keeping in mind that (a) the data for chimpanzees could come from provisioned or nonprovisioned groups, and that the former are much more aggressive than the latter (Horgan, 2010; Raffaele, 2010) and (b) their differences become less evident when they live in more similar environments (Boesch, 2002).

What Behaviors Do Chimpanzees and Bonobos Have in Common?

Chimpanzees and bonobos both may live in large groups of 120 or more with a fission-fusion structure, meaning that all the animals belong to the same group, but they are usually in smaller subgroups of changing animals at any one time. Both species are increasingly threatened by poachers, tourists, and farmers encroaching on their territories, which likely makes them more aggressive than they were in the past. Males of both species have dominance hierarchies. Chimps and bonobos both groom their friends' fur as well as their own. Males and females of both species have been seen now and then to capture prey and to eat meat. Meat is hard to come by, though, and is a tiny part of their diet—less than 2% weekly.

How Do They Differ?

Much more is known about chimpanzees than about bonobos because Japanese men studied bonobos and published their results first in Japanese, whereas Western researchers, often women, did work on chimpanzee behavior that received much media attention in the West.

Chimpanzee populations are far more numerous and spread out in Africa than are bonobo populations, which are restricted to small areas of the Congo. Chimps are somewhat heavier (males average 95 pounds and females 73 pounds) than bonobos (males average 86 pounds and females 68 pounds). Chimps have a squat, robust shape, but bonobos are slimmer—think Arnold Schwarzenegger compared to Rudolf Nureyev. In silhouette, a bonobo looks very like a lithe person. Chimpanzees have relatively longer arms than bonobos, but bonobos have longer legs.

Bonobos have canine teeth that are of similar size in males and females (as do humans), whereas in chimpanzees, the male canines are larger. Fossil skulls of the human ancestor *Australopithecus* are remarkably similar to those of bonobos, except that bonobos have a smaller brain case (Zihlman, Cronin, Cramer, & Sarich, 1978).

Bonobos, which are lighter in weight, spend more time in trees than do chimpanzees. Bonobo females have larger, more swollen breasts when nursing than do chimpanzee females.

Chimpanzees are far more aggressive than bonobos, with males known to have killed other males. Yet, although chimpanzees have been observed by thousands of researchers in scores of sites in Africa for more than 50 years, very few animals have been murdered. Only 31 intergroup killings have been recorded, of which 17 were of infants. Many groups have never been involved in killings. For any one research site, there is on average only one killing every 7 years (Horgan, 2010).

Experimentally, chimpanzees have exhibited altruism in sharing food both with kin and non-kin alike (Horner, Carter, Suchak, & de Waal, et al., 2011), but bonobos in the wild have done this much more commonly. Male and female chimpanzees sometimes kill infants, but bonobos do not. Annoyed chimpanzee males are likely to be tense and aggressive in their interactions, whereas irritated bonobos turn instead to heterosexual and homosexual sex. Chimpanzees use aggression to settle quarrels; bonobos use sex. Bonobo females, like humans but unlike chimpanzees, do not advertise estrus and very frequently—far more than chimpanzees—copulate regardless of the stage of their estrus cycle.

Bonobos and chimps both move about bipedally on occasion. Bonobos do so more naturally, usually when they are vigilant or carrying something; chimps usually do it to enhance their displays (Videan & McGrew, 2001).

Bonobo females get along splendidly, their closeness lubricated by sex; by contrast, most female chimps spend little time together. Bonobo females (which readily form coalitions) are dominant to males, whereas for chimpanzees, males are dominant. Bonobo mothers and their sons are much closer throughout their lives than are mother and son chimpanzees.

Bonobos are more sensitive than chimpanzees. During World War II, bombardment of the city of Munich caused all three bonobos at the zoo to die of heart failure, but the chimpanzees survived (de Waal, 1997).

Vanessa Woods (2010), who worked with orphaned bonobos and chimpanzees in African sanctuaries, found huge differences in the behavior of the two species. As one example, two bonobos shared food even if they did not know each other, but chimpanzees never did this. As another, chimpanzees were far more aggressive, whereas bonobos were friendly in the extreme once they got to know a person; this involved the person receiving a heartfelt "bonobo handshake" from their orphan friends that involved, to his or her dismay, either a penis or a clitoris.

The use of tins or pails reflects the temperaments of the two species. Mike, one of Goodall's males, saw empty kerosene tins as an instrument of aggression (Goodall, 1986). As a longtime inferior male in the dominance hierarchy, he decided to aim high. He picked up two empty tins by their handles, then sat down to stare at six superior males grooming each other and ignoring him. Slowly he psyched himself up until he stood erect and rushed flat-out toward these males, pant-hooting and banging the tins together, making a tremendous clatter. The males fled in panic. Mike had invented a new weapon that catapulted him into the alpha male position of his group—a position he held for more than 6 years.

By contrast, bonobos in the Yerkes Regional Primate Research Center in Atlanta, Georgia, were given a plastic pail used to carry food (Raffaele, 2010). Soon they had invented other uses for it: for holding water or urine, putting on their head, inverting for use as a seat, or playing with as a toy. Chimpanzee neighbors at the zoo that could see the bonobo innovations were also given pails to see what they would do. They used them only for violent displays, slamming them against the bars of their cage or kicking them across the floor.

In summary, chimpanzees and bonobos behave differently in many ways. If one were to choose which species had ancestors most like modern humans' progenitors considering the information listed here, certainly the species with the most humanlike characteristics would be the bonobo. If one wanted to prove that humans had a long history of aggression as they evolved, however, one would chose the chimpanzee, as was shown in chapter 3. This is what aggro-men have done. In their book *Demonic Males*, Richard Wrangham and Dale Peterson wrote that around 5 million years ago, "chimpanzee ancestors and human ancestors were indistinguishable" (p. 23). They also stated that "chimpanzee-like violence preceded and paved the way for human war, making modern humans the dazed survivors of a continuous, five-million-year habit of lethal aggression" (Wrangham and Peterson, 1996, p. 63).

There were many species of apes that became extinct and about whose behavior nothing is known, but to Wrangham and Peterson, the evidence was convincing that chimpanzee male behavior resembles that of humans' male ancestors because their belligerence seems a mirror image of humans' own, with "potentially lethal violence," "lethal raiding," and aggressive male coalitions front and center (p. 24). Because they saw human beings as a violent species, they chose the occasionally fierce chimpanzee as humans' nearest relative. These authors discussed bonobos at length but concluded that their lack of aggression makes them less like humans. Because chimpanzees and bonobos came from the same stock, these authors claimed that their joint ancestors were aggressive like their chimpanzee descendants but that bonobos largely lost their aggressive instincts during their evolution. Of course, there is no evidence for this at all. How could there be? Besides, it seems more than a bit ridiculous to know that bonobos and chimpanzees have very different behaviors when they have been separated for about 2 million years, yet to expect that humans' last common ancestors from more than 5 million years ago must have the same aggressive behavior as chimpanzees today.

Perhaps bonobos are found wanting by aggro-men for reasons beyond their good natures? Is it because female bonobos dominate the males? Is it because they are too sexy by half? (Frans de Waal [1998] reported that *GEO Magazine* planned to publish an illustrated account of bonobo social life but panicked when they realized that individuals have enjoyable sex perhaps once every hour.) Bonobo females are far more strong-minded than are most female primates. Males are larger than females, but they cannot boss them about because the females readily form feminist coalitions that can cause males grief. When females visit a feeding site, males turn aside or depart so they can keep their stalks of sugarcane to themselves. Male bonobos do not themselves form coalitions because they are bonded to their mothers. (This relationship can give them prestige in the community; if a mother has a high status, her son can parlay that into an increase in his own status, as described in chapter 8.) In addition, males cannot curtail sex for females because females are delighted to have sex with each other. Maybe aggro-men downgrade bonobos in part because they are such worthy feminists?

In fact, no other primate makes a good model for human behavior. Among the apes, gibbons come closest to humans in terms of their mating system (tight, long-term pair-bonding and nuclear families), and bonobos are like humans in their willingness to have sex outside of estrus and their use of it to avoid aggression and strengthen group cohesion (or pair bonds, in the case of humans). But humans are unlike bonobos in many ways, too: People mate in secret (and are, in fact, the only primates known to do so), are inclined to be jealous if their partners flirt, and either gender often becomes angry, even violent, at a partner's "extra-pair copulations," as the biologists delicately phrase it. We are not aware of any studies that compare the frequency and types of aggression in humans with those of other apes or other primates in general. Although this could be an interesting area of research, introspection might be more helpful to understand why some people are aggressive some of the time The explanation of biological or cultural roots of some types of human aggression will not be found in the behavior of other primates (Bekoff & Pierce, 2009).

We agree with aggro-men that for human beings, men work well together in coalitions (Tiger & Fox, 1971). Women do too. For proto-hominids such as *Australopithecus*, we envision a peaceable society of nuclear family nomads who did not make a practice of quarreling with their friends or seeking to destroy their neighbors. Over millions of years, they raised their children and had enough leisure to explore new lands, develop new customs and language, and create innovations that now help to define humanness.

ENDNOTES

1. The behavioral reports of other species such as Japanese macaques and rhesus macaques are also suspect if these species were provisioned during research projects, as was often the case. David Hill (1994) reported that the frequency of aggression between provisioned Japanese macaques while they were foraging was more than 20 times that recorded in smaller, natural groups.
2. Concealed ovulation is an important aspect of primate mating systems: in promiscuous species such as chimpanzees and baboons, the female advertises her estrus with visible swellings and other signs to attract males, but in monogamous species with no need to attract males, these signs are absent.
3. To put this into perspective, as mentioned previously, gorilla DNA orthologs (the coding part) is 98.5% identical to humans', and orangutan DNA is 97.0% identical (Goodman et al., 2005).

Chapter 5

Appeasement

Forestalling Aggression in Peaceable Primates

Although Robert Sussman is an anthropologist, much of his research is involved not with human beings but with monkeys and apes, a popular choice for many scholars in his discipline. After all, other primates are humans' nearest relatives, so their behavior could shed light on that of human groups. The evolutionary paradigm that is accepted generally for social species is that males compete with each other for the right to mate with females, whereas females may compete to ensure they have enough food; they must have nourishment to carry their young through pregnancy and nurse them as they grow. Some experts insist that "agonistic relationships are an especially important organizing feature in primate groups" (Sussman, Garber & Cheverud, 2005, p. 84). Other organizing principles center on the selfish gene theory, whereby the imperative of mating is the most important role of an adult, and competition theory, wherein the most aggressive animals prosper. (Scrappy individuals may prosper in their love lives, but statistics show that males of such competitive species will die relatively sooner than those of muriquis, for example, which never fight each other [Bronikowski, Altmann, Brockman, Cords, Fedigan, Pusey, & Stoinski, et al., 2011].)

Because of such theories, much primate research has been devoted to combat and other aggressive behaviors either within a group or between groups. It is far more exciting to record animals attacking each other than animals loafing about doing nothing much, so there has been little research on nonaggression. Sussman and his colleagues, Paul Garber and Jim Cheverud (2005), decided to find out exactly how different species of primates *did* spend their days. For each of 81 research papers involving 60 species, they tabulated how much time on average during the day monkeys or apes spent in social interactions, how much of this time was agonistic, and how much was friendly. Time spent feeding or traveling or sleeping at night was not included. Friendly behavior included grooming or playing with another, sharing food, and huddling and alliance formation of two or more individuals. Agonistic interactions were fighting, visual or vocal threats, submissive gestures, and evidence of one animal displacing another lower in the dominance hierarchy—so that much of what was deemed agonistic behavior was not even combative. More recent research confirms these findings (Sussman & Garber, 2011).

The chart Sussman and his colleagues constructed shows that individuals of all species spend far more time interacting in friendly rather than in aggressive ways. For 10 species of New World monkeys, congenial behavior accounted for 86% of all social reactions. For seven species of Old World monkeys, the average figure was 85%, and for lowland gorillas it was 96%. Obviously, nearly all interactions are friendly, and these form the basis of individual social bonds. All societies are held together not by fighting, but by cooperation and compatibility. Within a group, individuals may fight now and then, but these struggles do not take long and the combatants are not usually hurt: They live on to squabble again at a later date and often to reconcile (de Waal, 1989). Whereas many primatologists feed into the concept of male monkeys and apes being competitive and aggressive, Sussman and his colleagues believed, as we do, that they are by nature sociable, with occasional fits of pique.

Appeasement Behaviors

In this chapter, we discuss behaviors that have evolved in various species to dampen or eliminate within-group aggression and thus retain generally peaceful and therefore healthy societies. Males are by far the more aggressive sex; thanks to their lusty libidos, virtually all females in estrus become pregnant sooner or later, so there is far less need for them to fight among themselves. Here are many of the behaviors that help mitigate aggression in primate groups.

Dominance Hierarchy

A great deal of attention was given in the past to primate hierarchies. The largest animals in a group are usually the most aggressive and most dominant, and the smallest and least aggressive are lowest on the totem pole; age is sometimes important as well. Observers often make these distinctions by tossing a bit of food between two animals to see which one grabs it first and is therefore the more dominant.[1] If females are smaller than males, which is usually the case, then the males tend to be the dominant sex. Males and females may have separate dominance hierarchies or they may share the same one. For some species of leaf monkeys with multifemale–single male groups, females organize themselves in hierarchies and leave the male out of them.

Depending on the species, dominance rankings may be strictly observed (as in Japanese macaques) or largely ignored (as in bonnet macaques). The hierarchy of bonnet males means little in the course of daily life. During greetings, dominant males mount subordinate ones or vice versa. Either may groom the other. Unlike Japanese macaques, bonnet males often play with infants, huddle in groups when at rest, and are in general much more individualistic in their behavior (Silk, 1994).

Dominance hierarchies are common and important in primate and other social species because they ensure that harmony reigns most of the time within a group. Otherwise, members might be constantly squabbling over various resources. Although primates fight in order to rise to a higher ranking that delivers more privileges (such as the best-quality food and resting places), once the ranking is stabilized aggression is reduced dramatically. Every individual knows who is below and who is above him or her in ranking and will generally submit to those above rather than fight for what is wanted, thus not wasting time and effort and perhaps being injured in the process. In chimpanzee groups, for example, skirmishes may be common but serious fights are rare (Goodall, 1986). Some primate hierarchies are so stable that an alpha male may remain in the position for many years, with little fighting on his part: Chimpanzee Ntologi from Mahale Mountain M group was the dominant male from 1979 to 1995 (Jahme, 2001, p. 134). Robert Sapolsky (2001), who studied olive baboons in the field, reported that the elderly Solomon as an alpha male was so ferocious that he was able to quell any baboon that stepped out of line with a glance or at most a swat. His troop mates were intimidated enough by him that none seriously challenged his reign for an entire year. (Sapolski was terrified of him, too, after Solomon knocked him off a rock, shattering his binoculars.) Jane Goodall's chimpanzee Mike, once he had attained alpha status by loudly banging tin cans together and terrifying his group mates, held on to this position for 6 years.

Physical Touch

Touching is a fundamental social behavior that both recognizes and reinforces bonds between individuals and within a group. When chimpanzees are excited, which often sets off aggression, they make physical contact with a companion, touching, embracing, kissing, or mounting each other (defined as a close embrace from behind, sometimes accompanied by pelvic thrusting motions) (Goodall, 1986). These activities, along with various vocal noises, also calm an animal that is startled by a nearby skirmish or presented with an abundance of food. Male bonobos are less excitable than chimpanzees, greeting, reassuring one another, and resolving conflicts by mounting each other or touching rumps (Furuichi & Ihobe, 1994). In Tibetan macaques, two males may lick each other's penis, which seems to ease tension between them (Ogawa, 2006). Among muriquis in Brazil, daily life in the treetops involves frequent embraces and reassuring touches among all members of a troop (Strier, 1992).

Some behaviors that deter aggression may have evolved as an offshoot of social grooming. Many primate species show a dramatic increase in body contact between former opponents as they reconcile following a conflict.

A unique category of touching used by spider monkeys is grappling. Males are more gregarious than females, going about in larger subgroups and more likely to do things together such as grooming and patrolling the boundary of their territory. Individuals seldom fight each other, but still there is an underlying rivalry among them about who mates with whom (Aureli & Schaffner, 2008). This unease has been connected with an unusual behavior called grappling that is thought to counter aggression. It begins with a series of tense approaches and retreats by a nervous young male toward a mature male. Finally, the two interact with a combination of embraces, wrapping of tails together, touching of faces, and sometimes manipulation of each other's genitals. These various types of body contacts are carried out for many minutes between the two. Young males are highly attracted to older males but are also highly uncertain about their own place in the group. Grappling eases the tension between the two animals and presumably lessens the possibility of future aggression.

Reconciliation

Frans de Waal has written a whole book, *Peacemaking Among Primates* (1989), about the reconciliation of monkeys and apes in captivity. Individuals that assault others may make amends by grooming or hugging them, activities that reduce the negative impact of their attack and perhaps prevent a counterattack later on. If a bonobo grabs or slaps another, an offense that takes only a second or two, the assailant often quickly offers reassurance to the other in the form of sex or an embrace (de Waal, 1989).

Some species are far more likely to be conciliatory than others; reconciliation among the less hyperactive bonobos is more common than among chimpanzees, for example, even though bonobo squabbles are less fierce than chimpanzees' are. In the zoo, de Waal's chimpanzees sought reconciliation after a conflict about 40% of the time; by contrast, reconciliation among wild chimpanzees after a spat occurred on average about 17% of the time (Wittig, 2010), and much sooner than for the captive animals—about 2 minutes after the squabble. It might seem that reconciliation is a product of the reasoning human mind, but it has obviously been in place for many millions of years.

Sex

The expression "kiss and make up" is true. A subset of reconciliation is sex to resolve social tensions. Humans do it. Chimpanzees do it. Bonobos restore and maintain peace by having sex among all possible combinations of individuals. Females have sex with females by rubbing their clitorises together and have sex with males often facing stomach to stomach. Males have sex with each other, sometimes dueling with erect penises, and females have sex with youngsters, showing how it is done. Unlike most species, bonobos do not restrict sexual activity to the period in which a female is in estrus.

Social Grooming

One individual grooming another, allogrooming, is huge in decreasing tension and aggression. This reciprocal activity is so common in primates (which have agile fingers that can reach most parts of their own bodies) as to be a defining feature of human beings' order (Dunbar, 1991). Indeed, "picking through the fur of another individual is perhaps the most characteristic of all primate behavior" and has been well documented for more than a century (Sparks, 1967, p. 150). In a survey of 44 species of free-living primates, all but five groomed other individuals as well as themselves (Dunbar, 1991). At its most fundamental, grooming is hygienic;

it removes parasites such as ticks, fleas, lice, and leeches and may have allowed early primates to flourish in tropical habitats.

For chimpanzees, social grooming occurs in many contexts, such as after a separation, reconciliation after a spat, and alleviation of fear, but also for reassurance in times of excitement, which is when aggression often takes place (Goodall, 1986). It can then calm an individual that might otherwise have started a fight. By contrast, male bonobos that do not form coalitions seldom groom each other (Furuichi & Ihobe, 1994). For a variety of species, aggression in a social group is reduced during periods when there is increased grooming (see Barrett, Gaynor, & Henzi, 2002).

Over the eons and probably long before the evolution of humans' ape ancestors, grooming had evolved into a social function that goes well beyond the need for hygiene. What about human beings? In many cultures, individuals do groom each other, such as brushing another's hair or braiding it into cornrows. Adult white people are little involved in this, although a wife may remove lint from her husband's suit or smooth down his hair. Robin Dunbar (1996) suggested that the evolution of language has eliminated the need for social grooming that tends to build bonds between individuals. One can only groom and be groomed by so many partners in a day and still have time for feeding, resting, and other activities. When the group size of human ancestors became too large for the social function of grooming, language took up the slack in maintaining friendly relationships.

Hormone Levels

Robert Sapolsky's (2001) research with wild olive baboons revealed complex relationships between hormone levels and social behavior such as aggression. Individuals that groomed others most had the lowest stress hormone levels. Dominant males had low levels of stress hormones during stable years, but their levels rose during unstable times when fights broke out. A type-A individual that was, by definition, all too anxious to get into scraps, had twice the resting stress hormone level as an easygoing male. Low-ranking males had elevated stress levels all the time, reflective of their difficult lives: Their immune systems functioned less well than those of more dominant baboons, and they had less of the good version of cholesterol and apparently elevated blood pressure. Sapolsky was pleased to report that obnoxious subadult males had the highest levels of testosterone, so it is not possible to equate dominant males plus testosterone with aggression.

Presenting

This is a gesture used in many primate species in which an individual turns its rump toward another's face, sometimes to offset aggression. In chimpanzees, it is done by a submissive individual, usually a female toward a male as if seeking copulation. The presenting animal that hopes to prevent an altercation may have been threatened or may simply be passing by a scary individual (Goodall, 1986). In olive baboons, the gesture is rather like a greeting, although if a female presents to a male he may sniff her bottom to see if she is sexually receptive. Male baboons present to other males if they chance to meet, as do females and infants (Strum, 1987). In Tibetan macaques, a subordinate male approaching a dominant one presents his hindquarters to him before exhibiting his penis by raising his hind leg, as if he were urinating. His superior accepts this homage by lightly touching the lesser monkey's penis or buttocks, a response that increases the younger male's feeling of confidence as he settles down beside his elder. The tolerance shown by dominant Tibetan macaques toward lesser males is thought to help prevent these males from forming revolutionary coalitions against them (Berman, Ionica, & Li, 2007).

Behavioral Displays

A huge silverback gorilla rose to his feet and began to rhythmically beat on his chest with his fists; sharp *pok-pok-pok* sounds reverberated through the dense vegetation. Dian Fossey (1983) was glimpsing her first gorillas. Later, Fossey watched a confrontation near her camp when the young silverback leader Uncle Bert of Group 4 encroached on the territory of experienced Beethoven's Group 5. Beethoven did not like this. He strutted down the hill to within 4 feet of Uncle Bert. The two males adopted a rigid stance parallel to each other, their faces averted. The members of their groups fell silent, awed by the tension. Nobody moved. Finally, unable to stand the strain any longer, Uncle Bert rose on his hind legs and beat his chest, then slapped down the vegetation between him and his rival. Beethoven, enraged, charged Uncle Bert, who fled ignominiously, followed by his group, all screaming hysterically. Beethoven did not deign to pursue them but strutted back uphill. The following day, there were further confrontations, but eventually Uncle Bert's group withdrew from the area. Had the two males chosen to fight, one of them might have been killed, and both would probably have been wounded. It was much better instead just to confront each other with displays of aggression.

In the last few years, several researchers have written about aggressive displays by silverback males toward females in their group; these have a different context than displays against "enemies." Male gorillas are much larger than females, which indicates from an evolutionary perspective that in general large males are more successful in mating than smaller ones, and that there is competition among males. An aggressive male's display toward a female conveys to her the information both that he is strong and can protect her and that he is strong and can hurt her or her infant. This anomaly is discussed in chapter 8.

Chimpanzees, which are much more excitable than gorillas, have many display behaviors that reflect aggression. These include threatening with their arms, swaying branches, throwing rocks (usually with poor aim), swaggering on hind legs with arms akimbo, and charging an opponent (Goodall, 1986). Such actions, which are often intensified by barks or screams, may be directed in frustration at an inanimate object such as a tree rather than at an offending colleague. Jane Goodall noticed that high-ranking males displayed more than did their inferiors, which tended to stop displaying entirely if a dominant male approached them.

In 2010 a thorough study was published of male aggressive displays in ursine colobus monkeys living in the forests of Ghana, the first of its kind. It involved 26 males that, on various occasions, gave three kinds of displays: (a) "stiff legs," when a monkey held his legs straight out from a tree branch for varying lengths of time; (b) "jump displays," in which an individual leaped through the canopy with legs held stiffly; and (c) bouts of loud calls (not considered here because of the difficulty of knowing which animals responded to the calls). Researchers Julie Teichroeb and Pascale Sicotte (2010) followed the males for many months in the forests, observing who displayed to whom and with what result. These two odd exhibitions required a fair amount of energy to perform—for example, the duration of "stiff-leg" exhibits decreased with an increasing number of bouts performed—so they must have some useful purpose (Sicotte 2010, p. 278).

One hypothesis was that perhaps females watching these exhibitions were coerced into having sex. This seemed not to be so because females willingly mated with a number of males when they were in estrus. Perhaps females preferred to mate with males that gave many displays? This seemed to be partly true. But the displays of males overwhelmingly were directed at individual males and other groups. The displays served in part as threats, communicating the displaying males' strength so that others would not be tempted to attack them.

For an example of stylized male dominance displays in people and females' reactions to them, watch the classic film *West Side Story* and listen to its lyrics, or go to any high school during recess.

Agonistic Buffering

As its name implies, agonistic buffering minimizes possible aggression. This behavior was first described by John Deag and John Crook (1971) from Great Britain, who in the 1960s studied the behavior of Barbary macaques living in Morocco (see chapter 7). These monkeys are not aggressive by nature, but their contact with infants helps keep them that way. In this species—indeed, in many of the macaque, baboon, and gelada clans—infants are sacred to both males and females (would that it were always so in human beings), and picking one up may avert an attack.

Here is an example: A young male with an infant clutched to his abdomen sprints on three legs toward an older male. When he arrives, he presents the baby to this senior in one of several ways. He sits down, placing the baby between him and the male. Or, he may have the baby on his back, in which case, he stands before the older male while giving a lip-smack or teeth-chatter face (both appeasement gestures); he then pulls the infant off him to set it between them. Or, he stands diagonally to the older male, which mounts him, mouthing the baby on his back and pulling it toward him. Then the younger male grooms the second male. Such activities are contained within the group because agonistic buffering never occurs between members of different groups.

Appeasement gestures are so common and effective in Barbary macaques that all males feel free to remain in a group for as long as they wish, which is lucky for adolescents; in many species (for example, for most colobines), male youngsters are driven out of their natal group when they reach puberty. Having to roam about by themselves or with other bachelors in a suboptimal habitat where there are predators is dangerous. Because infants are so highly prized, perhaps the widespread handling of them helps to build friendship between two Barbary males and stability within the group. But at the same time, this contact with infants is stressful for the young males, causing increased levels of stress hormones (such as glucocorticoids) in their blood (Henkel, Heistermann, & Fischer, 2010).

Recently, scientists have proposed a satisfying evolutionary rationale for Barbary macaque male behavior (Henkel et al., 2010). Because agonistic buffering is important in this species in preventing or muting aggression among males, it allows young carrier males access to male networks. This access helps facilitate their links with adult males. Out of these links come coalitions, and males in coalitions have higher reproductive success than other males. Therefore, a young male that has the nerve to pick up an infant and approach a scary older male may be stressed out, but he has a brighter reproductive future than a more timid friend.

Agonistic buffering is common in Barbary macaques but less so in other species, where it occurs in different forms. Are there different reasons for it in different species? Is there a general theory for this odd activity? Zoologist Andreas Paul and two colleagues (2000), who studied agonistic buffering in general, found it present in 11 Old World species, including baboons and macaques; with so many species involved, it could hardly be an anomaly, which implies it has an underlying evolutionary cause.

Obviously an infant being carried by group members is not in danger of infanticide, but this seems too slight and simplistic a reason for the handling. Stretching their imaginations (a phrase we use not to indicate disbelief but to emphasize the tentative nature of hypotheses that attempt to explain the evolutionary basis of observed behavior), the researchers derived other hypotheses. Perhaps a young male carries an infant to a male that is its father, in which case the older male will accept it and its carrier without aggression. In that way, the carrier may be perceived as a colleague. Or, the infant may be familiar to the recipient, perhaps a nephew or niece carrying some of his genes. Perhaps during a tense situation, a presenter carries his own infant to another male to indicate he is ready to defend himself with maximum force if attacked. In theory, the other male would then react with friendliness rather than aggression. However, sometimes the carrier cannot be the father, either because he is new to the group or because he is too young.

Another idea is that if a male attacks an infant, there will be massive retaliation against him by the mother, her relatives, and perhaps some males. The infant therefore is used as an implicit threat—any harm to it will cause mass attack from the group. Or, perhaps an infant being carried about acts as a passport to discourage aggression among males or to facilitate males forming coalitions within a group. Will the infant so presented to a dominant male have an easier subsequent entry into a coalition? Certainly, agonistic buffering is rare in Japanese and rhesus monkey troops, where male coalitions are also rare. By contrast, in Barbary macaques, agonistic buffering and male coalitions are both very common. The matter needs further study that will be simplified with the recent discovery that paternity can be determined from DNA samples from feces.

Bridging

Bridging behavior is related to agonistic buffering but is even stranger. One cold January morning in Huangshan, China, the large male Tibetan macaque Youshizhi was leaving the provisioning site after eating his fill of corn (Ogawa, 2006). He swung himself into a tree where many of his pals were perched on branches. As he passed Zhiming, a 1-year-old infant, the son of Zhi, he swept him onto his stomach. His mother did not mind this, and neither did Zhiming, who clung to his large friend. Youshizhi moved on along the branch until he was facing Duyanlong, another adult male readily identified because of his deformed eye. Youshizhi turned and presented Zhiming to Duyanlong, holding the infant by his shoulders. Duyanlong took hold of Zhiming's legs and lifted them up so that Zhiming, with his back facing down, formed a bridge between the two males. Duyanlong put his head between Zhiming's legs and sucked his penis while Youshizhi freed one hand so that he could touch the penis too. After three seconds, Duyanlong released his hold on the infant and Youshizhi clasped him again to his stomach. Soon the two adults were sitting companionably side by side, with Zhiming wandering away from them along the branch. Youshizhi went after him, brought him back, and performed another bridge with Zhiming held between him and Duyanlong. This time, the bridge ended with Zhiming clinging to Duyanlong's belly. Youshizhi left them to climb to the top of the cliff above them. Duyanlong began to groom Zhiming and then to groom another infant, too. Later all three of them followed Youshizhi up the cliff and out of sight.

When Hideshi Ogawa first saw this bridging behavior, he was amazed, especially because it was so common, occurring hundreds of times within a few months. Monkeys do not usually hand things to each other, so how could it have evolved? Why would one monkey offer a baby to another in that way? Male baboons often grab an infant to hold in front of them if they are about to be attacked or when they are in a tense situation, but the Tibetan macaques used infants when they were relaxed and in a friendly mood. Often youngsters are playing together when a male chooses one to hold and then to offer elsewhere as a bridge.

Ogawa found that it was frequently a subordinate male that instigated bridging by carrying an infant to a more dominant male, and these males were often friends that liked to groom each other and sit together. Bridging seemed to function as a sort of calming activity and to greet and gratify other males. The infants were a tool to this end; the younger the infant, the more potent the tool, although even 3-year-olds were sometimes used. Ogawa noted that bridging seems to mean that "a child is the pledge of the group" (Ogawa 2006, p. 66) for Tibetan monkeys, just as the Japanese proverb states that "a child is the bond between husband and wife" for human beings.

Did the males have particular favorites? Yes: Youshizhi and Sanxiong, for example, frequently chose Zhiming, making him their "favorite infant" and the two adults Zhiming's "familiar males." A familiar male and a favorite infant were extremely close, with the infant sometimes sleeping at night on his male's belly or the two grooming each other. There was no evidence that these two were closely related, such as brothers or uncle and nephew. Ogawa found that a male carrying an infant often took the infant for bridging to one of the infant's familiar males in a "my friend's friend is my friend" relationship. Male Tibetan monkeys are astute, taking the infant not to any male but to one of his familiar males, which would presumably be pleased at this choice.

People have positive behaviors that help to cement social relationships. When a person in Western society is introduced to a stranger, he or she will usually smile and shake hands. Arabs will do so and touch their hearts; Japanese bow. After the greetings, a person thinks of something to say that might interest the other person: Do you work at this place too? What a lovely day it is. Such mundane exchanges seem to be comparable among male Tibetan macaques to bridging activity.

Determined Pacifism

In general, if two primates are going for the same food or resting place, the subordinate usually backs off to avoid a confrontation, knowing that he or she will likely lose in any competition. If there is a fight, the winner takes all—retaining or gaining a higher place in the dominance hierarchy that leads to a better life. The odd thing about hamadryas baboons is that they do not have a dominance hierarchy; if two males are likely to cross paths, they both back off. Hans Kummer (1995) sometimes saw two hamadryas baboon families feeding in a fruiting tree; if the males came upon each other suddenly, both families rushed away so the fruit remained untouched. Similarly, if a squabble between two bands became too heated, perhaps over the use of a cliff face for sleeping, all the animals fled, leaving the cliff unoccupied. No group ended up losing, but neither was victorious, either.

Why does this species not have a dominance hierarchy like most other primates? These baboons have a unique social system in which a male chooses one or several females when they are very young to live with him for the rest of his life, or as long as he can keep them for himself. If one of these females does not stay near him, he punishes her with a neck bite. This unit has been called a harem, although sometimes a female does sneak away to have sex with another male. If adult males began fighting with each other, they would have to leave their females undefended. One of them might be badly wounded, given the sharp canines of baboons, and the females would flee from the commotion, probably never to be rounded up again by either combatant. As Kummer explained, property owners have nothing to gain by a fight with another male, but everything to lose should all his females leave; the chances of regaining them would be slim.

In a way, this arrangement is sad because it means that young males that have grown up together abruptly stop being friends when they choose their first mate, even though they still belong to the same band. As youngsters, all the small males play enthusiastically with each other, and when they are adolescents, they sometimes form a gang that hangs out and sleeps together on a ledge apart from their birth families. As young adults, they prefer to groom each other rather than other individuals. But when they enter what amounts to a marriage, their erstwhile friends no longer hang around. Life is now serious. For human beings, many a newly married man misses his erstwhile friends and they him, for a while, but the pleasures of marriage usually soon outweigh the lost camaraderie. Were this not so, how could humans remain in nuclear families?

Kummer recounted an enlightening story about the mind-set of hamadryas males. Once, when his group of researchers was transporting animals from one site to another, they had to put two big males from the same band together in a large cage during a long road trip. When the first male saw that another was to be put in his cage, he turned away without looking at him. The second entered the cage tentatively, then also sat down in his corner with his back to the other. They both seemed to realize what a dangerous situation this was: Should their eyes meet, they might feel forced to fight. Yet each also put great trust in the other: While his own back was turned, each was vulnerable to a surprise attack that would certainly do great damage. Both tensely ignored the other, though, and the trip ended safely.

Chasing

For both squirrel monkeys and Nilgiri langurs, when one troop meets another, there is frantic chasing, with males tearing through the treetops with their penises erect. It is surmised that such activity dissipates tension and aggression, with nobody getting hurt (see chapter 12). Silvered lutung males chase too, in the same circumstance (Bernstein, 1968), although the state of their penises during these encounters has not been observed.

Related Behaviors

Many behaviors are not strictly appeasement but also function to reduce aggression. Following are some examples.

Subterfuge

Higher primates are smart enough to be able to undermine aggression with subterfuge. Richard Byrne and Andrew Whiten (1988) reported that a baboon being chased by another may suddenly stop short and adopt an alert posture, as if suddenly detecting a predator. His pursuer will immediately stop too, to look around for the danger. With luck for the first baboon, the quarrel between the two may be forgotten.

Fission-Fusion Groupings

Chimpanzees, bonobos, and spider monkeys all have fission-fusion societies (Aureli & Schaffner, 2008). In these, as noted earlier, members of a large community often break up into smaller groups of varying composition that go separately in search of food or water or refuge. This is important where these commodities are limited; should too many individuals converge on a fruiting tree or a small pond at once, fighting would certainly break out. However, it is impossible to document if aggression is reduced because of different animals choosing one group rather than another with which to associate.

Sperm Competition

For species in which a variety of males mate with each female in estrus, males have evolved large testes with a huge number of spermatozoa. During copulation, a large amount of semen is released into a female's vagina, where it is joined, should the female soon mate again, with the semen of her next consort. In effect, sperm from any number of copulations will each try (metaphorically) to be the one to fertilize the egg that will develop into the female's next infant (Birkhead, 2000). Because it benefits each male to have as many sperm as possible, these species have evolved large testes. Female chimpanzees, for example, often have semen dribbling from their vaginas after a mating orgy. There are other aspects to sperm competition, both physiological and behavioral. Having sperm competition within the female's reproductive tract may reduce physical competition between the males that are mating with her, but there is no way to prove this.

HUMAN APPEASEMENT OF AGGRESSION

Because the weight of testes as a ratio of body weight is more than 4 times less in men than in chimpanzees (Harcourt, Harvey, Larson, & Short, 1981), men have not evolved to copulate with many women. This situation is similar to that of other apes in nuclear families, such as gibbons, as well as gorillas with their harem mating system and many other mammals. What are the appeasement behaviors for angry human beings? A man may use some of the same methods as do social monkeys and apes:

- a dominance hierarchy—he may curse at his wife or a working-class friend when he is irate, but not his boss;
- physical touch—he may hug a toddler having a tantrum or rest a restraining hand on a furious adult;[2]
- reconciliation—he may have "make-up sex" after a spat (the best kind of sex, according to comedian Jerry Seinfeld);
- display behavior—he may shake a fist rather than smack the face of an aggressor;
- determined pacifism—he may grit his teeth and be polite to a lying political opponent; and
- fission-fusion—he may change his membership to a rival dart club to avoid a cheating member.

To mitigate aggression, people neither carry infants about to impress their betters nor compete with sperm (thank goodness), but they do use gestures such as shrugging, standing tall, or hanging the head, and, of course, they use speech. Human faces are also able to express a variety of appeasement expressions with 44 facial muscles, twice as many as the chimpanzee has and more than any other mammal (Walter, 2006).

In summary, the possibility of being aggressive is present in the genes of every primate individual, but species have evolved a multitude of ways in which aggression is largely kept under control within groups so that their members can lead productive and largely peaceful lives (Fry, 2006). Human beings continue to use many appeasement gestures that were developed by their ancestors.

ENDNOTES

1. Donald Sade (2000, p. 237), who studied rhesus monkeys in the wild, successfully started fights among the animals "by presenting a small piece of platano and observing which monkey displayed threat and which displayed submission, irrespective of which got the treat." Later, he came to abhor this intrusive technique because it fostered aggression within the group.
2. Research has shown that higher-status people in offices and classroom settings touch subordinates more often than vice versa (e.g., Leffler et al., 1982).

Section II

Primate Dyads

The hundreds of monkey and ape species on earth have evolved myriad different behaviors that enable them to live successfully in their specific environments. Many of them display characteristics that are similar to those of human beings. The next four chapters focus on dyads: relationships between males that live in monogamous relationships with females, together raising their young; males that are attracted to infants that may or may not be their own; males that form sturdy nonsexual liaisons with females; and males that become close friends with other males. All of these relationships also occur in human societies. Aggro-men note that aggression is a feature of primates in general, which it is. These other behaviors are features of primates, too, so human beings could as readily have inherited them. It will never be known exactly how evolutionary behavior in primates has unfolded, but human beings certainly have traits like those that are discussed in the next four chapters.

Chapter 6

Nuclear Families

Working as a Team

"Nuclear families"—a term we use rather than "monogamy," following Fuentes (1998)—are important to a discussion of aggression because they obviate the need for it except in limited circumstances. Members of nonhuman nuclear families do not assault each other, and the males do not usually fight other males except in defense of territory or resources.

While we limit our use of "monogamous" to describe human and other hominoid mating systems for reasons mentioned in Chapter 2, mammalogists nowadays are comfortable with the term when describing long term, stable pair bonds between an adult male and an adult female. Recognizing the frequency of extra-pair copulations (EPC) in most monogamous mammalian species that have been studied so far, at frequencies that range from miniscule to nearly universal, biologists distinguish "social monogamy", that is, living in a stable pair bond even though there might be EPC, from "genetic monogamy" when there is no DNA or other evidence of mixed paternity. Many studies have now shown that EPC rarely affects stable pair bonds in mammals.[1]

Although they are the rarest type of social organizations among primates, nuclear families have evolved independently in all major primate lineages (Kappeler & van Schaik, 2002). In Madagascar, diurnal lemurs are polygamous, but most nocturnal ones live in small nuclear groups (Fietz, 1999). Some tarsiers live in nuclear families consisting of adults and their offspring (Cowan, 2006). In South America, individual groups of tamarins and marmosets may oscillate among monogamy, polygyny, and polyandry (Dunbar, 1995). Many New World monkeys, including titi monkeys (Valeggia, Mendoza, Fernandez-Duque, Mason & Lasley, 1999), howler monkeys (Bolin, 1981), and owl monkeys (Fernandez-Duque, 2007), have nuclear families, whereas for apes, these are present only in humans and gibbons. Nuclear families take many forms in these groups and have evolved for various reasons (as reviewed later), but they are generally characterized by male parental care of young, the small size of testes,[2] and low levels of aggression.

As was seen from fossil evidence in chapter 2, clues in the bones of human ancestors show that at least as far back as *Ardipithecus*, 4.4 million years ago, they likely lived in pair-bonded nuclear families within small social groups. The nuclear family works fine for infants because both of their parents are devoted to their well-being. It works well for females, too; maternity is never a question, and they have a partner who helps to rear them. At first glance, it might not seem to work as well for the males, who cannot be sure that the young they help to raise do indeed have their genes. Birds are particularly suspect because they can "cheat" in only a few seconds with a "cloacal kiss"; primate copulations take somewhat longer. But monogamy is all its variations obviously has adaptive value, or it would not have evolved in so many species, nor persisted so long in our own lineage. What the male loses in parental uncertainty, he gains in high survival rates for his offspring because both parents are involved in their care.

The nuclear family amounts to a collegial team effort, with parents working together to produce and raise healthy offspring. It shall be seen for tamarins and marmosets, small New World monkeys, that mothers suckle and look after their infants when they are very young, but as they grow bigger, the fathers take over their care, lugging them about on their backs and finding tasty insects for their delectation. Because of this help, a mother can more quickly become pregnant with her next litter.

There is no place or reason in this scenario for aggression between the parents. Indeed, nuclear families may have two or even more males becoming part of the group, as shall be seen. The all-important female mates with each of them, so they all have an interest in helping to raise her young, which could be their own. These monkeys get along well together as they pursue their common aim. Presumably if a male became recalcitrant, he would leave to try to better his possibility of being a genetic father. These are friendly-father societies.

Why Nuclear Families?

Here are seven main hypotheses as to how nuclear families evolved in mammals and why they did so in primates especially (Reichard, 2003a; Schülke, 2005).

Resource Defense Hypothesis

A food resource is small but valuable enough that a nuclear family can maintain complete control of it. Gibbons, for example, defend a territory that contains a sufficient number and variety of fruiting trees to sustain their group; both males and females have enlarged canines to fight off attackers.

Capital Investment Hypothesis

A beaver couple has an extensive labor investment in their dam and lodge that either would lose if they left the area. Likewise, humans build permanent habitations for their families, the investment in which is a strong inducement to monogamy[3] (and for modern-day human pairs, because divorce settlements often make separation difficult).

Long-Term Relationships

Nuclear families and the paternal care of infants are part of the same hormonal, physiological, and psychological package. Two factors are pertinent here: First, the longer a couple has been together, the better able they are to raise their progeny successfully. In many or most mammals and birds, experienced parents have more and healthier offspring than do neophytes. This may keep primate couples together for the long term. (For large birds such as migratory Canada geese, couples that have been together for several years have the best reproductive success. Time is of the essence, because young that are hatched in the north in the spring must be strong enough to fly south at the end of the summer. Each parent knows just what to do to make sure this happens—build the nest, copulate, lay eggs, oversee the hatchlings, and defend them from predators as they grow. Efficiency may mean the difference between success and failure.) Secondly, for human beings, delayed maturity and longevity make paternal investment in children important.

Infanticide Avoidance Hypothesis and Predation Avoidance Hypothesis

Maybe nuclear families evolved to protect the young. The attendance near an infant of a male as well as the mother may prevent infanticide in some polygamous species of langurs. The need for guardianship against infanticide has been promulgated for various species such as gibbons, but as shall be seen in chapter 10, infanticide does not occur in nonhuman primates living in nuclear families. Preventing predation is a vital and primary function of males in many species, whether they are in polygamous or monogamous mating systems.

Female Dispersion Hypothesis

To obtain enough food in an area of scarcity, females may be so spread out that a male cannot feel sure he will find a mate or, if he does, that he alone mates with her. He may decide[4] to stay with her in a nuclear family relationship rather than try, and fail, to keep adequate company with two females. Indeed, if the females are very dispersed, he may be unable to find a mate when mating season comes *unless* he stays with one.

Paternal Care Important for Infant Survival

This is true for gibbons, humans, at least some lemurs, and some New World monkeys, including tamarins and marmosets, as discussed in the following section.

Intersexual Feeding Competition

Oliver Schülke (2005) proposed that fork-marked lemur females accept a single male to share their territory as a defense against feeding competition from other males whose ranges unsystematically overlap theirs. There is not enough food available for a male to consort with more than one female.

Obviously, monogamy is not a "one size fits all" mating system, and many species have evolved it, or lost it because of the exigencies of their habitats, as well as their biology and evolutionary history. It is not all about competition for resources. If each environment had one optimal mating system, then one should see all the primates in a place with the same mating system, but this is not the case. In Borneo, for example, along one stretch of riverine forest, one can see monogamous gibbons, solitary orangutans, two species of harem-mating macaques, and two genera of leaf monkeys, each with a different mating system (Harding, 2011a; 2011b; 2011c). Variety, it seems, is the spice of life.

Here we continue our survey of diversity in primate pair-living, with a look at an exclusively South-American group, the tamarins and marmosets.

Tamarins and Marmosets

Tamarin and marmoset (callitrichid) males are the most devoted fathers of all. These tiny squirrel-like primates from South American rain forests have a unique lifestyle, with nine different species all practicing intensive male caregiving—some tamarin fathers spend up to 78% of each day with the babies (Whitten, 1986). The males in these nuclear families usually father twins rather than a singleton; they carry these twins soon after they are born, and for many weeks they work hard to supply them with choice insects and fruits to eat. Each mother is devoted to the twins too, especially physiologically—together, they weigh about 20% of her body weight at birth, and as they grow, she supplies them with not only milk but also solid foods. (If women had newborn babies that were relatively as heavy, their twins would weigh 14 pounds each.) Callitrichid infants are obviously a huge drain on the energies of their mother (Tardif, Santos, Baker, Van Elsacker, Feistner, Kleiman, & Ruiz-Miranda, 2002).

Why are callitrichid males such admirable fathers? The evolutionary theory goes like this (Pook, 1984): Tamarins and marmosets could have evolved to their diminutive size for two possible reasons. It could be that small size and multiple births were primitive traits retained over millions of years, or it could be that these characteristics evolved over time. This second possibility is the more likely one. Presumably, tamarin and marmoset ancestors were evolving to occupy the ecological niche of small individuals eating fruit and insects that had been filled in the Old World by rodent species that did not then live in the New World tropics. Yet, the smaller the callitrichids were, the more they had to fear from predators. If the species was going to survive, the animals had to have a high rate of reproduction.

This was a challenge for the females. How could they increase their production of babies? In general, the smaller a female mammal, the larger in proportion to her weight is her newborn. Fairly small monkeys in captivity have difficulty giving birth to a single young that weighs 15% of the mother's weight, so how could an even smaller mother manage? The best way to evolve to a smaller size, with adults weighing a pound or less in some species, was to have twins (nonidentical) with heads that could more easily slip out of the mother's birth canal.

So far, this makes sense. But how could the mother raise twins successfully by herself, especially when they were so hefty compared to her own weight? She could not, so it made sense for their father to help bring them up. Because they were a monogamous pair, he had as much invested in them as she did and would do what he could for them. As has been seen, his task was to carry the young about when they were a few weeks old and, along with their mother, help find for them high-quality food such as large insects to eat.

Australian Anne Goldizen (1990, 2003), who studied the behavior of saddle-back tamarins in Peru, knew all this before she began her field research, so she was puzzled by some of her findings. To travel with two growing youngsters on one's back is hard work, especially when it is necessary to cover as much as 2 kilometers a day through a tropical forest hunting for food, so Goldizen assumed that the father would be the one doing this. However, she noticed that when an encumbered father became tired and scraped the twins off his back onto a large branch, it was not the mother that hurried to the screaming infants but another male. He too carried them about as solicitously as the first male had done. What was going on? These tamarins lived in small groups, so could this male also have mated with the mother and be the possible father?

Female callitrichids do indeed mate with two males if possible, or even three or four, all of which form presumably a happy nuclear family. The males will later help rear her youngsters even though only one of them can be the father. If there are offspring from an earlier litter in the group, they will help too. This unusual situation, called cooperative polyandry, produces youngsters that are likely to live to maturity. By contrast, tamarin twins raised entirely by two parents rarely survive. Male members of this group have evolved to be not only excellent fathers but also good friends with each other as they labor to rear as many young as the female can produce. The teamwork of these little monkeys makes for maximum reproductive success.

SIAMANG GIBBONS

As Harding made his way at dawn along a trail in the mountains of Malaysia, he began to hear a deep "whoop, whoop, whoop" far off in the forest. The path brought him closer to the sounds that got louder. Suddenly, a cacophony of calls erupted that seemed to be made by two or three species. The "whoops" continued, but now there was a "yeow!...yeow!" that accelerated in timing, seeming to reach a climax. Then Harding heard a third call, a single yell, like a protracted "YAAAAAAAAAA...." This was two troops of siamangs yelling at each other across the mutual border of their territories. After a few minutes, the sounds separated, one troop fading into the jungle and the other coming closer, continuing its cries. The gibbons crossed the trail, swinging from branch to branch high up in the canopy. There were two adults and two juveniles. Occasionally, they all stopped and the adults made the "yeow...yeow" followed by the "YAAAAAAAAAA..." call, the latter accompanied by much leaping about and shaking of branches.

Siamang gibbons are intensely family oriented, with as many as four youngsters of different ages living with their parents—young born every 2 or 3 years are not fully independent until they are 7 or 8 years old. Every day is family day; in one study, family members were nearly always within 10 meters of each other, were involved in the same activity three quarters of the time, and at night slept in trees close together (Gittins & Raemaekers, 1980).

All gibbons live in forests, but it is only the male siamang, the largest of the "lesser apes," which has the same behavioral pattern as the tiny callitrichid males, of being closely involved with young offspring (Dielen-theis, Zaiss, & Geissmann, 1991; Gittins & Raemaekers, 1980; Whitten, 1986). Life is tricky for tree dwellers: They may be eaten by predators such as snakes or birds of prey, or they may be injured or killed in falling to the ground. For their safety, infants must be carried at all times. The mother siamang looks after and transports her single young about for its first 10 or so months. After that, as nursing winds down, she increasingly spends less time with him or her, and the male or males of the group take up the slack. Rarely, the female may live with two males that get along well together and both mate with her. Thus, either may be the infant's father (Lappan, 2008). Less frequently (in 12% of female copulations in one study [Reichard & Sommer, 1997]), she lives with her mate but has extrapair copulations with neighboring males. Many gibbon species have yet to be studied in detail (Bartlett, 2011).

Field research of siamang life is labor intensive. In Sumatra, Susan Lappan from South Korea and three assistants carried out extensive follows from one sleeping tree to to the next so that they could record all activity between five infants and the males belonging to their five groups. The males sometimes groomed and played with the infants, but 98% of interactions involved the infants being conveyed by the males from one tree branch to another during arboreal travel among food sources and to sleeping trees (Chivers, 1974). By the end of the second year, the youngster is sure-footed and sure-handed enough to travel on his or her own, except for needing help at the more difficult crossings over open areas. They sleep in the night tree with either their mother or a father.

In siamangs, male care is not additional to maternal care, because the more care the males provide the young, the less is offered to these young by their mothers. That makes sense in a way because a mother's physiological resources can then be diverted to benefit her next fetus. It seems that this infant is born sooner to a female that has had more male help in carrying her young. What does not seem to make sense is that in two-male groups, both males take turns carrying their infant, but they still together carry him or her for less time than did single males of a nuclear family. And, the males that were so helpful with the young spend extra time with the mother, although they might have been expected to do this hoping for more sex in the future—if that was their motive. But, as was described earlier, some of the same hormones and neuronal reward-system pathways involved in family bonding are also involved in bonding between fathers and infants. Maybe evolution has simply programmed gibbon males to care for infants because they will pass on more genes if their offspring survive than if they do not, regardless of how much sex is given for the effort. In fact, there is no evidence that increased paternal attention is associated with increased copulations, and gibbons as a whole copulate infrequently—observers can watch for hundreds of hours without seeing it (Brockelman, 2009; Lappan, 2008; Palombit, 1995; Reichard & Sommer, 1997; Zhou, Wei, Li, Lok, & Wang, 2008).

All gibbon pairs sing duets, which reinforce the family bonds (serving the same function as sex in humans). The male and female partners in a mated pair sing different parts for their duet, usually given at dawn. Each species of gibbon has its own highly stylized call. The duets and other calls serve not only to bond the pair but also to enable them to keep track of each other in dense vegetation and to signal to neighboring gibbons that their territory is occupied by a feisty family. In the territorial sense, duets and loud calls[5] take the place of displays and potential aggression between groups at the territorial boundaries of other primate species.

The siamang duet, a complicated composition that would confound even seasoned opera singers, goes like this: The male gives two deep booms, the female then booms once, the male booms twice again,

> and the female must immediately come in with accelerating high-frequency barks. After about the fifth bark, the male should utter an ascending boom, the female's barks should speed up, and the male should do a bitonal scream. At once the female must start another series of barks, and after five the male should do a scream, this time a ululating scream. The female does some fast high barks, then both of them bark and hurtle about. ((McCarthy, 2004, p. 113)

The Louisiana Purchase Gardens and Zoo in the United States recorded how a siamang male and female formed a bond when an 18-year-old male that had lived alone for 12 years after his partner died was housed with a 6-year-old female. It was love at first sight (Maples et al., 1989). After they were put into the same enclosure at the zoo, the two siamangs began to practice this duet right away, their performance improving quickly. During the first few months, the percentage of started great-call sequences that were completed correctly rose from 24% to 79%, until finally almost all were perfect.

> One of the most common mistakes was for the female to start her first set of accelerating high barks without waiting for the male's second double boom. The other common mistake was for the male to bungle his bitonal scream, giving either the ululating scream or the locomotion call instead. (Maples et al., 1989, p. 114)

Do males get something else from monogamous relationships besides exclusive access to sex with their mate—love, perhaps, or the satisfaction of fatherhood? As recompense for limiting their mating opportunities, do females derive a benefit from their males besides help in rearing their offspring—love, possibly? Certainly family members live a harmonious lifestyle day after day, year after year, with only occasional squabbles among themselves or with neighbors about territorial boundaries. It is odd that although all monogamous species have friendly relationships within their families, there is little written about this and much written about aggression and violence among primates in general.

White-Handed Gibbons

Before beginning his research on white-handed gibbons in Thailand, Thad Bartlett (2003) believed from reading the literature on this species that social interactions between groups were almost exclusively aggressive: Fierce territoriality was thought to be at the heart of gibbon behavior. Whenever two groups met at their joint territorial boundary, the male gibbons gave loud calls and shook branches in aggressive displays, attracting human voyeurs to the area with their uproar.

What Bartlett saw in the wild, though, was often quite different. Most meetings between groups were hostile, but friendly encounters between neighbors took place on a regular basis. Not only did juveniles from two groups play together, but once a male groomed a juvenile from another group for nearly 20 minutes. Such encounters in the past had been overlooked by researchers because they took place in silence and were seen only when the animals involved were used to people following them about and comfortable in their presence.

CRESTED GIBBONS

On his first morning in Vietnam, Harding stepped out of the guest house at Cuc Phuong National Park just before dawn as the jungle awakened. Green magpies and treepies were making their raucous racket, bulbuls called musically, warblers warbled, babblers babbled, and laughing thrushes laughed. Suddenly, from the jungle just below the guest house, unmistakable primate calls rang out, drowning the birdsong with their tenor and intensity. These were from crested gibbons of several species housed at the Endangered Primate Research Centre. Harding listened, amazed at the variety of the dawn duets that were sung by the mated pairs. Some langurs housed there added to the general cacophony. Then, from high up in the mountains behind the park headquarters, came a reply: Wild white-cheeked crested gibbons were answering their captive relatives.

Crested gibbons are not well known to Western scientists because of the lack of access, until the last couple of decades, to their ranges in Vietnam, China, and Laos, and because poaching has driven all seven species close to the brink of extinction. They now remain only in the remotest mountains of Vietnam, Cambodia, and southern China, although healthy populations persist in northern Laos. In 2009, a species thought to be extinct was rediscovered along the border between Vietnam and China, and in 2010, a new species was described from central Vietnam.

Because of environmental circumstances, these apes may have unusual lifestyles. Like other gibbons, crested gibbons are mainly monogamous and territorial, with the female rearing few offspring to adulthood—perhaps not even half a dozen—in her lifetime and both male and female subadults dispersing (or being driven out) from their natal territory at around 5 years of age. The very first report, by French naturalist Jean Delacour (1933), suggested that white-cheeked crested gibbons may sometimes live in polygamous groups because he often saw two or three buff-colored females in groups of otherwise black males and subadults.

Other groupings have been observed in black crested gibbon groups by Chinese scholars. From countless hours of watching during several years in the Wuliang Mountains of Yunnan, Peng-fei Fan and his colleagues documented an instance of polyandry similar to that noted previously for tamarins (Fan, Jiang, Liu & Luo, 2006). In that case, a female lived with two males, mating with both. Jiang Zhou and colleagues (2008), after 29 months of close observation of a relict group on Hainan Island, reported a polygynous group: one male living with two females. Both teams ascribed these unusual coupling combinations to small numbers of gibbons living in highly fragmented habitats where it was not possible for pairs to find and defend their own territories with sufficient resources for their survival.

Nuclear families are a feature of many species of primates and other animals. Within them, cooperation and altruism, rather than aggression, are valorized. This type of social organization is obviously successful in a variety of habitats, including that of human ancestors. The roles of primate fathers in nurturing their and others' offspring has been given short shrift in the scientific literature and is largely ignored in the popular press. We will discuss this in chapter 7.

ENDNOTES

1. Many studies (reviewed by Cherkas et al., 2004) show that more than 18% of women in most Western cultures have been unfaithful to a steady partner or husband, and men more so; the "nonpaternity rate"—at which children in families are biologically unrelated to the presumed father—ranges from about 1% to more than 30% depending on the sample community. Such variability suggests that although there may be an underlying genetic component, cultural factors dominate the prevalence of EPC in humans.
2. This fits the sperm competition hypothesis: because only one male mates with a female, he does not need many sperm to compete with those of other males and therefore has smaller sperm-producing organs. Chimpanzees have huge testicles compared to humans, gorillas' are smaller, and gibbons' are the smallest among the apes, relative to body weight.
3. All other great apes build only temporary sleeping platforms that they use for one or a few days. Gibbons are the only apes that build nothing.
4. "Decides" is evolutionary jargon. All males of a given species may simply follow an instinctive pattern that evolved because of the advantage of monogamy versus the alternatives in a given resource/terrain situation; however, some species have a mixed strategy where he does, actually, consciously decide on one strategy or the other.
5. "Loud call" or "great call" is a specific term, meaning a territorial call that is made, usually only by males, often at dawn. It is characteristic of many primates, such as most colobine monkeys, howler monkeys, and some gibbons.

Chapter 7

Males Befriending Infants

In some species of primates, males are much more involved with infants than are most men. In human cultures such as that of the !Kung, a gathering and hunting tribe in Botswana, men spend about 14% of their time interacting with infants, about the same amount as the most devoted fathers in Western societies. But in some industrial areas, men visit with their babies as little as 45 minutes a week. In a world survey of paternal relationships in 80 cultures, 20% of fathers were rarely or never near their infants (Whitten, 1986). (Of course, these figures for Western civilization may reflect the understandable desire of savvy men to avoid messy situations that women [per force] handle. Dagg's late husband Ian argued and pleaded for years that he could not change diapers because it would make him throw up. This excuse worked until Dagg broke her arm and was unable to cope by herself in the age of diaper pins. She was more delighted than he to find that he did not vomit as predicted, so that even after her arm healed she could call upon him again during emergencies.) Men are more likely to pay attention to their children as they grow older, become increasingly independent of their mothers, and learn to speak. It is then that boys and girls benefit hugely from interacting with their fathers. By contrast, nonhuman primate males interact with youngsters mostly when they are very small.

In this chapter, we highlight a sample of species in which males play an important role in the lives of infants. These males may be the infants' fathers, but often they are not. Primates in general (with the exception of those few that commit infanticide for ulterior motives, as discussed in chapters 10 and 11) have infants that are highly vulnerable at birth yet are never harmed by members of their community despite the curiosity and interest they arouse; adults seem to realize that youngsters must be valued because they represent their future. Depending on the species, males may carry them about, cuddle them, rest beside them, or play with them. Youngsters of many species like to play with a larger male, and such play itself is a cooperative exercise. If one partner plays roughly, spoiling the fun, the other will neither trust nor want to play with him or her again (Bekoff, 2006). Dogs, for example, perform a play bow by bending only their forelegs, staring up at their friend expectantly, and often wagging their tails and barking. This gesture signals that all the exciting bites, bumps, or chases that follow are for fun and not to be taken seriously. Primates have other signals to indicate play behavior. Lucky youngsters that often interact with adult males begin to learn what it means to be an adult and a father.

Infants are paramount in some species in defusing aggression, as shown in chapter 5. This symbolic use of infants held by adult males is so complex and nuanced that David Stein (1984b) wrote an entire book about it in one species called *The Sociobiology of Infant and Adult Male Baboons*; out of the nearly 1,000 interactions he observed between adult males and infants, in none of them was an infant harmed.

The species that are discussed in this chapter represent a sample of the many ways polygamous males interact with infants. It has already been seen that fathers in species with nuclear families are attentive to their young.

Howler monkeys, which live in trees, as do most monogamous primates (humans being the obvious exception), hang out in small subgroups and have young that are treated with fondness by the males. Barbary macaques are of special interest because the concept of males using infants in a friendly fashion as tools within the group for "social buffering" or "agonistic buffering" was first reported for them. The related stumptail macaque males are also friendly toward youngsters. Japanese macaque males (unlike females) willingly adopt weaned orphans, offering them support that may enable them to survive to adulthood. Rhesus macaque males have the same innate impulse to rescue infants in distress. Young hamadryas baboon males are infatuated with infants that they carry about and cuddle; when the males are fully adult, they adopt female infants about a year old which they rear to become their future partners. Yellow baboon males interact with infants, which, because of this, will later be amenable to being held by them for the purpose of agonistic buffering. Gorillas live in mixed groups with one dominant male around whom the youngest infants play; he also monitors the behavior of the older youngsters.

Howler Monkeys

Male primate interactions with infants are especially notable in arboreal species where infants must be carefully guarded at all times lest they fall to the ground or be snatched by predators. In her study of black howler monkeys in Guatemala and Belize, Inge Bolin (1981) noted that the males were very tolerant of the infants in the six groups she studied, each comprising from four to seven monkeys. She followed these groups day after day, from dawn until dusk, until she had put in 477 hours, concentrating on everything that was affecting the infant (or infants, if there were two). She carted around a tape recorder so that she could describe what was happening from minute to minute. If she lost track of a group, their incredibly loud howls would sometimes lead her to it again. Ultimately, she was glad that she had studied six groups instead of just one, because there was much variability in how males and infants got along together.

Although male and infant interactions in one group were few, those in another exceeded the number of interactions between an infant and its mother.

The adult males were usually friendly toward the infant or infants in their group, especially as the youngsters grew older; on only two occasions (out of 3,139 documented) did a male react peevishly—when two infants insisted on jumping repeatedly on his back. Sometimes a group female deposited her infant beside a male before climbing far out on a tree branch to forage. The male would cuddle it in his arms until its mother returned. Once, when a small whining infant seemed abandoned at the top of a fig tree, the group males hurriedly swung up and remained beside it until its mother returned. This behavior is reminiscent of that of the other species that will be discussed in the following sections, whose males have a much softer spot than do females for orphans and other infants in distress.

As howler monkey infants grow, they increasingly seek out the adult males. When a youngster approaches an adult, he may ignore it or he may respond by holding, hugging, or muzzling it. Sometimes a male allows a youngster to playfully explore parts of his body or to jump on his back or swing on his tail. Sometimes when a male and an infant form a close bond, they rest or sleep beside each other. In another population of howler monkeys nearby, an adult male often touched a year-old female and moved about with her held closely to his body. He was especially protective of her when a female spider monkey came up and tried to hold her.

BARBARY MACAQUES

Barbary macaques, the only African monkeys north of the Sahara (and the only macaques in Africa), were one of the earliest primate species to be studied in the wild. In the 1960s, many of them lived in the mountain forests of Morocco in groups of 12 to 30, each with both male and female adults. Conditions are harsh there, especially in winter when the snow is deep and there is nothing to eat but the bitter needles of cedar trees (Whitten, 1986).

All Barbary macaques love babies. The infants grow quickly, so they are soon able to wander about by themselves among group members. Should there be an emergency, such as a threat from a jackal or an eagle, an infant will be scooped up by the nearest adult and carried into safety in a tree. Infants are safe even when neighboring groups of macaques mingle together occasionally at a source of food or water.

Female Barbary macaques love to groom and play with their infants. Sometimes juvenile females (but not other mothers) convey one around, becoming a center of attention because of this. But the adult and especially subadult males are even more captivated. They pick up infants as young as 1 week old to carry against their stomachs. For an older infant, a male may back up to a mother and baby, lower his hind legs so that the infant can climb onto his back, and then set off with his rider. When it comes time to rest, he may cradle the infant in his lap, mouthing, licking, nuzzling, smelling, or manipulating the penis of his young charge. Sometimes the two sit side by side while the elder grooms his little friend for 20 minutes or more. Or, the caretaker may simply sit and stare intently at the youngster, chattering his teeth. As the infants grow, the males become less attracted to them, giving up all interest in them by the time they are a year old. Each youngster has one or two favorite males that interact with him or her, and male caregivers restrict their attentions to one or two specific infants. These interactions are so important that they likely contribute to infant survival; in one study, the only infant that died was one that had received no care from adult or subadult males (Taub, 1984). Females in estrus mate with all the group males, so probably the infant and its favorite males are not related, but they could be siblings. This baby-positive behavior by the adult males has been described under various names: "paternal care," "paternalistic care," "paternal-like behavior," and "uncle behavior" (Deag & Crook, 1971). However, because it is usually not known which male fathered which infant, a better term is "male care," which is similar to "mother care," without the nursing aspect.

Male care benefits both participants—the baby because he or she enjoys being held, carried about, groomed, and protected and the male because he can use the infant in social interactions with more-dominant males. This social buffering smooths out relationships between two males. In Barbary macaques, few friendly approaches are made by one male to another without an infant being present. If a subadult male carrying an infant is able to slip into the central core of the troop where the females and the dominant males hang out, it may seem to his seniors, if this happens often enough, that he belongs there, and they accept him accordingly, much to the young male's satisfaction.

STUMPTAIL MACAQUES

Male stumptail monkeys are not very tolerant of each other (Richter, Mevis, Malaivijitnond, Schülke, & Ostner, 2009), but they do have a soft spot for babies. They live in Southeast Asia—or most of them do. A troop comprising two adult males, nine adult females, 11 juveniles, and 7 infants was transported in 1974 to an island in Mexico. Alejandro Estrada (1984) decided to document the social life of these infants and juveniles each day for 10 months after the animals had been allowed to settle in for a while. He collected 337 hours of notes on their activities, many of which involved interactions with juvenile or adult males. The adult males spent a great deal of time with their young friends—which may or may not have been their own—giving special attention to the male young. Was this because they had a relatively large penis they could tweak? The adult females focused only on their own youngster.

Estrada gave delightful sketches of adult males and infants interacting: the male carrying the baby on his back or hanging onto his stomach; the two sitting side by side and playing; and the male grooming his young friend, manipulating its genitals, and touching its mouth or hand. Sometimes two males held an infant between them in a bridging manner similar to that perfected by Tibetan macaques. However, sometimes a male was annoyed with an infant, acting toward him or her in a negative way. Ninety-one percent of male-infant interactions were friendly, but the other 9% were mildly aggressive—involving a stare, a slap threat, or a nip. The mandate of the males seemed to be that of mothers: to share social experiences with the young, both positive and (rarely) negative, that would help them grow into healthy stumptail adults.

Japanese Macaques

Junichirô Itani (1959) was the first scientist to record fatherly care in monkeys. In May 1954, he wrote observation notes about the Japanese monkey male Syaraku and a female baby. These notes have been transcribed into English as follows:

> His infant tries to join [some playing] youngsters. Syaraku quickly goes to it, picks it up to his breast, climbs up to the grassy plot above the provisioning ground, sits down and begins to groom the infant. At 2.40pm, he stops grooming and stands up in a hulking manner. The infant jumps upon his loins. Syaraku climbs up to the bamboo thicket above, carrying his infant on his loins.
>
> A few days later, Syaraku allowed his small friend sitting beside him to eat peanuts without preventing or threatening her. (Itani 1959, p. 83)

In the Takasakiyama troop that Itani was watching, such behavior was common in the spring and summer between elite males and infants. In that season, the females were giving birth and preoccupied with their next baby, so the troop's leaders and subleaders, any of which could be an infant's father, took over the care for those 1-year-olds and sometimes 2-year-olds that were at loose ends. The older ones that they hugged, played with, and protected tended to be weaker and smaller than average; perhaps because of this, the males continued their relationships with them longer than usual—not an example of survival of the fittest.

These friendships had other advantages than seasonal care for the infants, because they continued to a lesser extent as the infant grew. Occasionally later on, the two might wrestle together in play or sit near each other. The males enlarged the scope of their charge, although sometimes a youngster in the care of an alpha male became spoiled, bossing about other animals that could not retaliate for fear of its nearby alpha friend. The male might show his young sidekick a new food to enjoy, and in at least one case, the infant returned the favor, which it must have learned from another monkey.

Such friendly behavior by top males was unknown in many of the other Japanese monkey troops in the area, so it was a cultural rather than a genetic phenomenon. It would be passed from one generation to the next because the chosen infants were the ones most likely to belong to favored females that spent the most time in the central area of the troop along with the leaders. The favored male infants would in time pass on this behavior to the next generation. Because the central area was where the most powerful males hung out, ambitious adolescents were anxious to be there too. One smart young male, Uzen, realized that if he carried an infant on his "bosom" (Itani 1959, p. 62) he might have access there, which is what happened. The infant was, in effect, a passport to the haunts of the elite; Uzen would likely go far in the future. Itani concluded from his research that male care has a social value, even though it is not present in every troop.

The 100 Japanese monkeys of another troop, the Takagoyama group, living 50 kilometers south of Tokyo, learned about human vagaries the hard way (Hasegawa & Hiraiwa, 1980). When local people set up a monkey park in 1960 to attract visitors to their area, they had offered the monkeys lots of free food. But over the years, tourist interest slumped. In 1976, the park was closed and the feeding stopped; there were no longer enough human visitors to make the operation feasible. The monkeys were left to fend for themselves. But now there were many more of them, thanks to the free food, and fewer foraging sites because a new golf course and road had been built nearby and a gravel pit enlarged. The monkeys were forced to sneak into farmers' fields to grab something to eat. Many of them were shot or captured by farmers—so many that 40 mothers disappeared, leaving behind 40 orphans.

What could be done with 40 orphans? Here is what happened to Sable, who was about 8 months old and probably not fully weaned when her mother, Yaji, was killed. Researchers Toshikazu Hasegawa and Mariko Hiraiwa (1980), who were following the troop around in 1978, noticed that tiny Sable was in turn following the two highest ranking males, Bana and Sasuke. When the troop halted, she fed near the males, where she was protected and groomed by them. Soon she was adopted completely by Bana.

When the troop moved on, Bana put his hindquarters near her and patted her neck to signal her to climb aboard, which she did. He then carried her across open fields and roads to new feeding areas. For several months, he was devoted to Sable, grooming her carefully at least 27 times. Sable groomed him twice. The researchers did not know if Sable could have been Bana's daughter, because during the mating season females copulate with a number of males, but over a 4-year period, Bana befriended four other infants as well as Sable.

Nursing infants younger than Sable will die if their mother is killed, but those her age or older will turn to adult males instead of older siblings for protection and care. The adult females of the troop, with the odd exception, had no interest at all in looking after the orphans. Most of the orphaned youngsters in the troop younger than 2 years of age were cared for by adult males, as well as some juveniles whose mothers were busy looking after their next newborn. The males groomed their favored infants as much as mothers groomed their own young, and they let them ride on their backs longer than did the mothers, so they provided care at least equal to that of the mothers.

RHESUS MACAQUES

For rhesus monkeys in Nepal, there have been several incidents of would-be orphan adoption by males, although in these cases, the young were newborns that needed milk to survive (Taylor et al., 1978). Rhesus males are far less baby-friendly than are Japanese macaque males. They seldom interact to cuddle, groom, or play with infants, some even threatening or slapping youngsters that try to climb over them. But the orphan-rescue impulse is alive in them, too.

Spock was a high-ranking male in a troop of 120 rhesus monkeys; he did not have much time for infants. But one day in May 1975, researchers saw a juvenile carrying a 1-day-old male infant. Because there was no female in the area, he was probably Merlin, whose mother had died the day before. When the juvenile dropped Merlin so that he rolled down a grassy slope, Spock ran to retrieve him. Several females crowded around to examine Merlin, but Spock shooed them away. He himself became his guardian, carrying and grooming him gently at the edge of the troop where the two of them would not be disturbed. When Merlin began to squirm and cry, though, Spock became irritated, placing him on the ground and resting his foot on the tiny body. He turned Merlin over and tried to stop the crying by mouthing and shaking him, which did not help. When he picked him up again, like an inexperienced mother he carried him in the crook of his arm or upside down. But no matter how Merlin was treated, without food (which Spock did not offer him) or milk, he would die of starvation, which happened the following day.

The next year, this time in June, the same thing happened again. Spock adopted a second newborn orphan, this time a female whom he treated in the same way. The baby died of starvation on her third day, with Spock carrying her body around and grooming it for another day. Shortly after the deaths of the two infants, the researchers observed Spock for several more full days to determine how he behaved under normal conditions. They found that when he had had the babies, he had spent more time in quiet activities such as gently grooming them, had foraged less for food for himself, and had stayed farther away from the other monkeys in the troop.

Hamadryas Baboons

The infant-positive impulse occurs not only in macaques but also in baboons. Hamadryas baboon males are all attracted to infants, but their story is quite different from any that have been considered so far (Kummer, 1967). Subadult males have more maternal-type behavior than do juvenile males of other species. As soon as they are sexually mature, they often catch and cuddle very young infants which they carry about for half an hour or so. Motherless infants are invariably also swept up and "adopted" by these males. A young subadult male is frightened of adult males, so when he is near one, he may grab an infant, embrace it, and turn away from the adult. Sometimes he lowers his hind legs so that an infant or juvenile can climb onto his back and have a free ride. On one occasion, a 5-year-old subadult under stress carried off a large 4-year-old on his back. Such contact with youngsters lowers the stress level of the conflicted subadult.

Females never interact with infants in this way, and the motivation for it in young males disappears as soon as one attracts his own first female to be his permanent partner. This female is also very young, but she is treated differently because the male wants to teach her to stay near and follow only him. He does this by giving her neck bites whenever she wanders away. Such beginning relationships are very common; in one population in Ethiopia, they made up about one fifth of all one-male groups.

Yellow Baboons

Both hamadryas and yellow male baboons are infatuated with infants, but where subadult hamadryas males seem to enjoy infants for their own sake, yellow baboon males seem to have ulterior motives. David Stein (1984a,

1984b) studied the question of adult male–infant interactions for a year in a troop of about 45 yellow baboons living in the Amboseli National Park in Kenya. This Alto Group included about 12 infants and 16 adult males, the subjects of his research. After a young was born, its relationship with the various males was documented in 20-minute focal samples: Were a specific male and a specific infant within 5 feet of one another? Were they touching each other? Were they in extensive body contact? Stein was able to collect nearly 500 hours of data relating to these questions.

Males are eager to be involved with newly born infants during their first 6 weeks of life, embracing them, letting them ride on their backs or cling to their stomachs, grasping an arm or leg, or grooming the tiny creatures. After that, their interest wanes somewhat, although it may revive when the youngsters are 5 or 6 months old if there is a fight, at which time the males may hold them to their chest for agonistic buffering, as discussed in chapter 5. However, they still stay in touch in the long term. The males' decreasing interest in infants is compensated for in part by the increasingly mobile young, which begin to approach the males, wanting contact. When an infant is about a year old and his or her mother is ready to begin a new pregnancy, the infant seeks to spend much more time with adult males, even though he or she is sometimes harassed by other males for doing so.

Stein gave two probable reasons why adult males like to interact with infants. First, such youngsters make highly effective agonistic buffers. Second, by being handled by an adult male, an infant develops an attachment toward him that develops into a friendship. This infant when it is older will fully cooperate with the male in future agonistic buffering sessions. Infants that act as agonistic buffers probably gain benefits from the involved males, such as being defended from possible harassment and being allowed to forage near the male for the best food.

Infant baboons may be handled by a number of different males, but often both males and infants have "preferred" friends, meaning that they are more likely to interact with each other than they are with other individuals. Does this mean that a preferred male may be an infant's father?

Stein believed so, although this was impossible to prove at the time when he carried out his research.

Mountain Gorillas

David Greer, an American researcher with Dian Fossey's team, probably spoke for all his fellow colleagues about his love for gorillas: "They're so big and furry and beautiful, and yet so peaceful" (as cited in Raffaele, 2010, p. 20).

Alpha males can be great fathers. After all, if a group has been together for a while, the silverback leader knows that the infants in the group are probably his. Picture Beethoven resting on the ground after a lengthy bout of feeding, rolling onto his back and glancing at his youngest offspring, 6-month-old Puck, who was playing happily on the big stomach of her mother, Effie, sprawled nearby. He reached over to lift Puck gently up by the scruff of her neck so that she hung above him, her body almost disappearing in his huge hand. Then he as lightly returned her to her mother. Similar peaceful scenarios occur again and again in the life of the group (Fossey, 1983, p. 64).

Beethoven cared enough about his youngsters to chide them if they needed this. One day when Fossey was taking photos of her gorilla friends as they rested, a then-older Puck persisted in pulling on the strap of her camera so that she could play with it. Fossey refused to give it to her. Puck finally gave up, plopped onto the foliage nearby, and sulked for most of an hour. Fossey, feeling bad, then gave her a *National Geographic* magazine to leaf through, the colored pages of which kept Puck's interest until the group roused itself to return to new feeding grounds. Immediately Puck ran at Fossey and slapped her hard with both hands. Beethoven, hearing the commotion, gave an annoyed grunt. Puck ignored him and again hit Fossey, who was lying down. Beethoven bounded over to Puck, grunting, to glare directly down at her with furrowed brow and pursed lips. Nonplussed, Puck turned away to follow the other gorillas, pouting. As

Fossey packed up several minutes later to leave too, Puck ran back again, halted beside her, and stood with a hand raised as if to whack her one last time. She paused for a minute, perhaps remembered Beethoven's anger, then scooted off again after the others.

When the gorilla Pablo was not yet 4 years old and still nursing from his mother, Lisa, she abandoned him—not only him, but all her family in her natal Group 5. Fossey (1983, p. 99) and her helpers believed that Beethoven or one of her half-brothers might be her father, so her leaving was not unexpected. This is what most gorilla females do when they reach adulthood. But leaving behind her son Pablo was a surprise—he was a cute little guy with a friendly manner, although he looked a bit crazy because of his crossed eyes. Amy Vedder (Weber and Vedder, 2001) was doing research on what gorillas eat at the time, and Pablo's solution when it started to rain was to march up to her and huddle by her side, his head under her arm to keep as dry as possible. Vedder was conflicted, needless to say. As a scientist, it was imperative that she not interfere with gorilla life. But as a person, she yearned to keep Pablo dry, warm, and safe from harm. What should she do? Her scientific self won out. From then on, she kept her arms tight to her side when Pablo was around, feeling bad that she could not cuddle him next to her (Weber & Vedder, 2001).

Vedder hoped that one of the females in the group would look after Pablo, but they all ignored him. Luckily, his father, Beethoven, took over his care, something that the researchers had never seen before. During rest periods, Beethoven had Pablo sit beside him, sheltering him with his huge body if it rained. At night, Pablo shared his nest where they slept together. Nearly a year later, Lisa and Pablo saw each other when their groups met briefly, but neither paid any attention to the other. Because of Beethoven's care, Pablo grew up to be a rambunctious youngster—for example, rushing up when having his photo taken to peer at his reflection in the camera lens—and eventually the esteemed alpha leader of 43 gorillas, the largest number ever seen in one group.

Later researchers realized that in general, if a female died when her infant was as yet unweaned, the "orphan" turned in desperation not to other females but to the alpha male, following him about like his shadow and sharing his nest at night. One youngster that lost his mother at the age of two, well before being weaned, survived to adulthood thanks to the care of a silverback male (Stewart, 2003).

HUMANS

Chapter 2 described how humans' early ancestors evolved a system of nuclear families in which males invested a large part of their lives in raising their children and caring for their mates. As Eduardo Fernandez-Duque and his colleagues (2009) noted,

> The existing evidence implies a common physiological substrate for both paternal behavior and pair-bonds. Vasopressin, and perhaps prolactin and testosterone, apparently underlie the endocrinological bases of paternal care, and neuroanatomical reward pathways may be involved in the formation of attachment bonds. (p. 115)

Paternal investment—that is, fatherhood—was necessary, not least because their large brains and delayed development meant that dependent infants, toddlers, and youths could not survive without the regular care of both parents, and perhaps grandparents, too. For more than 4 million years, men and protohuman males have been good fathers.

What is the case for fathers and their children in indigenous human societies before they became "tainted" by Western values? Anthropologists have reported variations, but fathers often have a great love for youngsters. For example, for the Tikopia people of Polynesia, studied by Raymond Firth (1957), fathers were expected to take their turn in looking after a child, which they did with apparent interest and pride. Child care was an obligation that a husband and wife shared. The anthropologist often saw a child in the early morning and evening in its father's arms, or between his knees, or toddling along beside him hand in hand. If the child wandered away and was frightened, he or she quickly returned to the father for comfort, seeking bodily contact.

Barry Hewlett and Shane MacFarlan (2010) studied paternal investment in children in a variety of hunter-gatherer societies and other small-scale cultures. They found that, as in tamarins, male care decreased female reproductive costs and increased the number of offspring by reducing the interbirth interval and, in some cultures where fathers cared for the young more intensively, by increasing infant survival and toddlers' rate of weight gain. Fathers regularly transported infants and young children, for example, among the Aka, who travel 5 to 15 kilometers per day. Among Martu Aborigine adolescents, Brooke Scelza (2009) suggested that although grandmothers and other women can provide child care in place of a missing mother, there is no substitute for a paternal investment that increases the social competitiveness of offspring, such as when children inherit resources controlled by males (e.g., land or cattle).

Among the nomadic Bedouin of the Syrian Desert—a vast, arid plain of eastern Jordan, southern Syria, western Iraq, and northwestern Saudi Arabia—Harding participated in innumerable coffee rituals where the youngest son of the host, whether a shepherd or a Sheikh, performed the honors. First, the guests would be invited to take off their shoes and sit on narrow rugs arranged in a square. The youth would then bring a steaming Arab coffee pot, or dallah, and one tiny cup with no handle to each guest. No one talked during this performance. When the coffee ceremony was complete, the meeting began and the youth brought sweet tea in individual glasses for refreshment, not ritual. The youth watched and refilled each glass before it became empty. The host pretended not to be watching his son closely to make sure he did it right, and the son watched his father out of the corner of his eye for a hint of approval or reproach. One can find analogues for this little scene in every society.

As another personal example of fatherhood, when Dagg (1978) was living with camel herders in Mauritania while carrying out research on camels, most evenings she saw men cuddling an infant or a toddler in their laps as they relaxed around the fire.

Patterns of male care of children in Western societies is so widely reported, complex, and variable that a review here is unnecessary. Here is just one vignette: Go to any ball park, arena, or skating rink in the early hours of the morning, after working hours, or on weekends, and there will be fathers coaching sports teams, providing guidance, exercise, and training for adulthood not only for their own offspring but for the youth of the community. Fathers are scout leaders, equipment managers, and drivers for the myriad athletic and cultural activities of childhood. That mothers share these duties—with considerable proportional variability among families and cultures—does not lessen one of the defining features of adult male behavior in Western society.

In summary, there are a variety of reasons why social primate males and infants become special friends. An infant may stay beside an adult male for protection, and while hanging out with this male, he or she learns how an adult functions in a social context. If a youngster is a male, there is a greater likelihood that he will be accepted into a male group when he matures. The adult male that interacts with a mother's infant may make her amenable to courtship in the future. It seems certain that in most multi-male-multifemale primate groups, no male knows if he has sired any of the group's offspring because the females are promiscuous (Chapais & Berman, 2004). It is heartwarming, therefore, that adult males of many species give much time, energy, and comfort to infants that are likely unrelated to them.

Chapter 8

Males and Females, Without the Sex

A friend of Dagg's who while in Africa had photographed a female giraffe, her baby, and a male giraffe nearby was delighted with this picture. "I'll call it *A Giraffe Family*," he exclaimed.

"There is no such thing," Dagg said.

He looked at her with bewilderment. "Of course there is," he insisted. "You have to have a father and a mother to have a baby."

It is amazing how humancentric in their thinking most people are. Of course there have to be a male and a female to produce an infant (usually, anyway), but it is now known that females as well as males among mammals are usually promiscuous. Rather than thinking of strictly monogamous species, which may not even exist, one should think of socially monogamous species with nuclear families as defined in chapter 6. Aside from that, in all mammals the mother nurses and raises her own young and may assist other mothers, but the father, if he is present at all, is far less likely to be involved with his offspring. A male accompanying one or more females and their young usually offers protection from other males and from predators, nothing more.

For animal groups in general, there is an obvious disconnect between males and females. To pass on their DNA, adult males must sire offspring in a situation where these young are likely to mature and mate themselves. Usually this means a male will copulate with as many females in estrus as possible. Sometimes males and females prefer different food sources or habitats, which also keeps them apart (and helps ensure that females have enough food to reproduce and rear their young effectively.) Some primate species do, however, have contented asexual friendships between individual males and females. Here, we discuss male-female friendships in olive baboons, Japanese monkeys, gorillas, and two orangutans. Then we consider species in which mothers and their sons are bonded, often strongly: bonobos, chimpanzees, and muriquis. The chapter ends with a discussion of the extreme rarity of forced copulations in primates, despite misinformation in books by aggro-men who refuse to accept this reality.

OLIVE BABOONS

Baboons, along with Japanese macaques, are probably the most aggressive monkeys, in large part because they live in big, mostly terrestrial groups that are vulnerable to predation. Male baboons can be especially combative—they have been described as fighting machines because of their powerful muscles, wide gape, and razor-sharp canines. Armed with such weapons, they can compete with each other for the good things in life: food, females, and the best sleeping spots. (The potential for attack was emphasized for Dagg when she once had a wild male baboon sitting on the hood of her tiny Ford Prefect car. He repeatedly opened his mouth wide, revealing formidable canine teeth.)

If baboons are so fierce, why are they in this book? They are included because they have another side, at least for olive baboons, that belies this fierceness. Adult males attempting to join a troop may attack its members, but those that are already in the troop usually become close friends with one or a few females and in this capacity serve as protectors for them and their infants against aggressive males (Strum, 1987).

Such a male was big Sumner, who was close friends with Peggy and with Constance, too. He was older than them, confident and gentle. The three companions often hung out together in a relaxed way, with the females' infants playing around them. Peggy's daughter Pebbles loved to use Sumner as a jungle gym, climbing up his fur to his muzzle, falling on his face, working her way through his thick hair, then sliding off his big belly. As she grew older, Sumner became a springboard to leap on or off. He never seemed to mind. Sumner's very presence offered security for his friends. If Sumner moved, Peggy, Constance, and their infants often followed him. They repaid his friendship with grooming. Sumner acted as a bodyguard for the youngsters because if other females wanted to fondle one, which they often did, they would check with him first to see if this was all right; if it was not forbidden, they turned their backsides, lip-smacking, and with a grunt began to gently touch the infant. Sociable males such as Sumner were the most successful with the females.

Barbara Smuts (1985) wrote a book about such male-female relationships called *Sex and Friendship in Baboons* —with the word "Friend" throughout spelled with a capital *F* to signify this special alliance. Smuts described another congenial couple: Plato approached Leda as she fed, with her son Laocoon clinging to her stomach. He sat down beside them, grunting and lip-smacking, then pulled Laocoon toward him and gathered him to his chest. Leda continued to feed, ignoring him. Laocoon dropped into Plato's lap, where he played quietly. Plato became sleepy, his eyelids drooping and then closing. Laocoon suddenly leaped up toward Plato's face to grab a clump of hair by his cheek, then fell back down, jerking Plato head as he went. Plato, without opening his eyes, curled one hand protectively around Laocoon's small body (p. 111).

How did Plato benefit by befriending Leda and Laocoon?

1. Because of his friendship with Leda, she was likely to mate with Plato when she next came into estrus.

2. If Laocoon were his son, which he may or may not have been, Plato by his presence was protecting him from aggression by other males, increasing the likelihood that the youngster would grow to pass on his own DNA in time.

3. Plato might on occasion use Laocoon as a buffer, holding him to his chest to prevent another male from assaulting him. Infants are sacred in most baboon societies, where other males will never harm one. Were they to do so, this action would become a public event, with massive retaliation against the attacker by the relatives and close associates of the infant.

4. Plato was able to relax completely in their company.

Friendships are, in effect, a system of social reciprocity. A male baboon newly joining a troop seems to be thinking, as far as his actions go, "To be successful in this troop, I will need a few female friends, several infant friends (preferably at least one tiny black infant), and some male allies." Weeks or months later, he can call in his dues, using an infant as a buffer against an aggressive male, mating with a female in estrus he has befriended, and forming an alliance with another male against a daunting enemy. The female and her young benefit by his protection as his friend.

JAPANESE MACAQUES

Mino and M-59 were good friends, Japanese monkeys living among 200 other monkeys in the Arashiyama B troop in Japan. M-59 was high ranking among the males, as was Mino among the females; the elite tend to stick together, just as they do in human society. Yukio Takahata (1982) collected behavioral data during five nonmating seasons from eight adult males 10 years or older in the troop and from the 84 females that were 4 1/2 years or more old. During each of these seasons, the researchers followed each male for a total of 10 hours, recording during this time every female that was within 3 meters of him. The idea was to determine if certain couples were especially likely to be near each other (as friends are likely to be). Grooming each other was not common, but if it occurred, as it did between Mino and M-59, it represented a specially trusting relationship; in each session, the groomee is in a defenseless position should the groomer turn nasty.

Mino and M-59 retained their friendship through all five nonmating seasons that were studied, but they were not joined at the hip. Each had other opposite-sex friends as well: Mino spent special time with four other males, and M-59 visited with 27 other females. (One of his other special friends was 27-year-old Momo, who was senile and unrelated to him, but still he often sat near and groomed her before she died of old age.) These numbers sound like a lot of togetherness, but in the huge troop, only 65 pairs (10% of the possible pairs) were considered to have these peculiar-proximate relations (PPRs). Takahata suggested that the connection between a couple should be considered a "psychological affinity" because they were not that often close together physically. Sometimes several females clustered around a male, making it seem that the females liked each other as well, but if the male died, these females immediately dispersed.

So far, these friendships sound rather like those of the non-kin baboons that were described earlier. However, a huge difference is that for the baboons, friend-companionship leads to mating when the female is in estrus, whereas the psychological bonds between Japanese monkey PPR couples usually preclude mating. If a female is in estrus, her friend keeps away from her so she mates with other males instead.

PPR friendships are not particularly common in Japanese monkeys, but they do remain an evolutionary puzzle. Because infanticide does not occur within a troop, a female does not need help to protect her young. However, having male friends could be of use to her now and then: (a) If she dies, a male friend would likely adopt her youngster; (b) if she is low-ranked, she can improve her status by sitting near a higher-ranked male; and (c) if she gets into a squabble with another female, her male PPR friend may intervene to her advantage.

It is hard, however, to see the adaptive advantage in a male becoming friends with many females with which, because of this friendship, he will not mate in the near future. Having many female PPR friends may enhance his leadership charisma, but it will not increase his reproductive success.

MOUNTAIN GORILLAS

Dian Fossey (1983) adored gorillas, fulfilling her dream when she came to live in the Virunga Mountains of Rwanda for 18 years, becoming friends with scores of gorillas before being murdered because of her passion in 1985. Two of the first gorillas she encountered in the cold, wet mountains were Rafiki, whom she greatly admired—"so dignified and commanding of respect" (Fossey 1983, p. 139)—and his mate Coco. Both individuals were in Group 8 along with their four sons.

They were both old by this time and had probably been together for many years. Rafiki was thought by Fossey to be about 50, with Coco perhaps a little older and no longer able to reproduce. Because there were no young gorillas in the group, Fossey found it easy to quickly habituate the members to her presence. When the group moved from one feeding area to another, Coco lagged behind so that Rafiki often stopped to wait for her. When they rested after feeding, an activity (if one can call it that) that consumed about 40% of each day, Coco often started grooming one of the males, picking out burrs, parasites, and dried skin from his fur. This led to the other males joining in, each grooming another in a chain of industrious black bodies. When night came, Rafiki and Coco often shared the same nest.

One day, Fossey was able to hide from view to watch these animals. They were all moving slowly uphill as they fed, although Coco had become disoriented and was heading away from and out of sight of the others. Rafiki suddenly looked up and gave a sharp call that Coco heard. She turned slowly toward the direction it had come from above her. Rafiki and the others sat down to wait for her. She slowly made her way up the hill, meandering about somewhat; when she spotted Rafiki above her, she immediately made straight for him. As she reached him, they exchanged low sounds of greeting before they embraced, looking straight into each other's eyes. She placed her arm over his back, he did the same to her, and murmuring to each other, they slowly walked up and over the hill, where Fossey could no longer see them.

On another occasion, Fossey met Coco when the males of Group 8 were in a state of frenzy, along with members of Group 4, which contained four young females. The males were flying from tree to tree, beating their chests, breaking branches, and calling loudly to attract attention to their displays. Coco was huddled behind a tree trunk, one hand tapping the top of her head and the other arm crossed over her chest. She wore a resigned expression, as if to say, "Boys will be boys." When she saw Fossey, she glanced at her calmly and gave a loud sigh. The two females sat together, where they were joined briefly now and then by Coco's son Peanuts, who rushed down from the fray to exchange embraces with his mother before heading back to the excitement.

Two months after this, Fossey arrived for a day of watching to find the parents absent and the four males playing together. She backtracked along the trail of the group to find that Rafiki and Coco had slept in connecting nests for two nights, but after that, there was no evidence of where they had gone. Fossey assumed that Coco had died because she was never seen again. Rafiki returned alone to the group, but without Coco's maternal presence, squabbling increased among the five males. Eventually, two left the group and two females joined it, one of which presented Rafiki with a daughter. He was content to spend his last days in a close family circle.

For a big gorilla such as Beethoven, with his own extended family, life was as good as it gets. In 1978, the silverback had four female adults and their youngsters, his offspring, living with him. Perhaps because the females had all joined him from different groups, they were not that fond of each other, but they doted on him. Each wanted to sit beside her mate. If they squabbled about this, annoying Beethoven, he would quiet them with an irritated grunt. Mountain gorillas are not particularly noisy animals, their most common sound being what researchers Bill Weber and Amy Vedder (2001) delicately called "the flatulent by-product of their vegetarian diet" (p. 121). However, sometimes they broke into song while feeding contentedly. First one would begin a low rumbling sound, breathing loudly in and out. Then others would join in for a mix of sopranos, tenors, baritones, and basses to create intertwining melodies rising and falling in a natural rhythm and lasting for several minutes. The human audience did not know what this music meant, but they enjoyed it "as unrestrained expressions of individual happiness and group harmony" (Weber and Vedder, p. 121). Over the years, the other females left the troop or died, until alpha female Effie was the only one left.

Beethoven was especially fond of Effie, with whom he had five young. During rest periods, they watched like many married couples as their boisterous offspring wrestled or played tag or king of the mountain around them. However, they had spousal tiffs too. Once Beethoven noticed Effie eating at a juicy blackberry bush (Weber & Vedder, 2001). Enviously, he rushed over to the bush, pushing aside Effie, who gave a scream and fled. But at other times, Effie refused Beethoven's mute demand to let him into her refuge from the rain. On the first occasion, Weber and Vedder were squashed under a huge *Hagenia* log along with Effie and two youngsters. Beethoven was standing out in the rain, hoping Effie would allow him to squeeze under the log too. Effie at first just pursed her lips; then she gave a dismissive series of sharp cough grunts, at which Beethoven turned away to find some other shelter.

Another time, when he had built a slapdash day nest that was anything but rainproof, he tried to edge into Effie's nest, where she was warmly ensconced with two of her offspring, all happily dozing. Effie refused to move over at first, but then she relented and let him squeeze in behind her, although most of his body was still exposed to the rain. When their son Icarus was grown, he remained with his parents rather than emigrating to another troop. Together, he and Beethoven protected their group from other gorillas when they were near and made bluff charges to terrorize tourists, which gave them bragging rights when they returned home.

So far, this chapter has considered two older gorilla couples whose devotion to each other was described by Dian Fossey herself. Sandy Harcourt (1979), Fossey's student researcher, carried out a 2-year quantitative analysis of adult male and female behavior in Beethoven's group 5 and Uncle Bert's group 4. Each group also had a younger black-back male (that was usually ignored by the females) and a few females with their young.

In both groups, the adult females preferred to spend time near the alpha male rather than with each other. They were most likely to attain this goal if they were in estrus or if they had a young infant— Beethoven and Uncle Bert liked infants, which they protected by their presence, and the infants became accustomed to their fathers. Females gave birth near the males, but if their infant died, they lost their favored position. However, the alpha male had to try to pay enough attention to each female to keep her from leaving his group to join another; because no gorilla troop is very large, the alpha male has a limited number of females to produce his babies. Fortunately, alphas can compensate for a small group size with longevity: Some remain the head of their group for 10 years or more.

When a huge male gorilla, twice the size of the females, stands up and thumps his chest, everyone pays attention. Investigators therefore often write about these aggressive displays, which also include vocal snorts called "pig grunts" or "cough grunts." Often a male uses these displays to threaten strange gorillas or human beings: He does not want them around. What do such displays mean when he uses them within his group?

From the foregoing, it is apparent that the alpha male and the females in a gorilla group get along fairly well together. Occasionally there is a clash between groups, but basically the animals live an idyllic existence (were it not for interfering people), traveling and feeding, resting at midday, traveling and feeding again in the afternoon, and then making night nests for sleeping. Nevertheless, one researcher, Pascale Sicotte (2002), published a paper entitled "The Function of Male Aggressive Displays Towards Females in Mountain Gorillas." Despite this title, in her report she wrote: "Females suffer no wounding in the course of these displays and do not appear to be particularly fearful of the male" (Sicotte 2002, 278). What is going on here? Is aggression aggressive if no one is harmed or fearful of the presumed aggressor's behavior?

Sicotte postulated four possible reasons for such displays by a male: (a) to impress the females with his fighting abilities, (b) to "decrease potential competitive inequities between females," (c) to provide the females with "an occasion to confirm their subordinance to the male," and (d) to have a short-term influence on mating (Sicotte 2002, 277). In her analysis, Sicotte decided that the first three possibilities are correct, but the last is not, because males did not display more than usual when females were in estrus. The first three, however, are problematic. The females are already aware that the male thumps his chest effectively; this is what he commonly does after the group has finished resting to make sure everyone knows that it is time to follow him and move on to new eating adventures. The second reason might prevent females from squabbling among each other, although this is not common behavior. The third sounds sexist—the females may respond, but they do so perhaps to show that they have heard the male. Describing these displays as acts of "aggression," contrary to the title of the paper, seems contrived.

Gorilla researcher Martha Robbins (2009) has mulled the fourth question further, of whether male aggressive displays toward females are a form of courtship or of sexual coercion. She could not make up her mind, even after observing the behavior of a mountain gorilla habituated troop in Uganda for 66 months over 5 1/2 years. (Some research programs are incredibly frustrating.)

Why might it be thought that a male coerces a female to have sex? Males are twice as big as females, so even their presence can be intimidating. A female could perceive a strong male as being a good protector for her and her young, but also as a possible threat should she choose to leave his group for another, which some do. Unlike in Sicotte's study, the two males did show more aggressive displays towards females that were in estrus than they did towards those that were cycling or nursing. (Could it be argued that their behavior was an odd sort of courtship?)

Why might it be thought that a male does *not* use sexual coercion? Coercive sex is very rare in nature. When females are in estrus, they want to copulate. Should a lusty male attack or frighten a female, he could scare her away. Male gorillas display toward females but do not physically harm them; it is now known what these displays mean.

The few researchers who have looked into this questions did not report that mating followed or was connected to acts of aggressive display by a male toward a female. In Robbins' (2009) study, for example, two adult females seldom mated and did not have young. If there were sexual aggression, one might expect fewer births in multimale than in one-male groups, where there are more aggressive displays, but this does not happen.

Will the meaning of these male displays ever be understood? It is not probable. Robbins noted how difficult the research is. Only a few groups have been studied long-term, and the variables are endless: How dominant and how old are the males in a group? What is their history and that of the females? Will a female want to mate more with an aggressive male?

Here is an alternate explanation. Although it is intimidating to human observers, perhaps the male's chest-thumping display is not aggressive at all. Maybe it is just like human males showing off in front of women with attention-getting loud talk, vigorous movements, and active gesticulation. Researchers are not supposed to be anthropomorphic, but was it not anthropomorphism that originally categorized male gorilla displays as aggressive? Because it is known that females do not respond fearfully to male "aggressive" displays, and that males and females in general get along well in their groups, this interpretation is at least as likely to be true as framing the display as aggression.

ORANGUTANS

Adult male and female orangutans are described as largely solitary, with the female providing the only infant care, as far as is known (Delgado & van Schaik, 2000; Schürmann & Van Hooff, 1986). Adult males, which are twice the size of adult females, may associate with subadult males, but not for companionship: Their interactions usually involve competition for a receptive female (Galdikas, 1985). At least, this is the conventional wisdom. Harding saw many orangutans in the wilds of Borneo, usually either solitary males or females with young. But once, in a semi-wild setting (a large protected primary forest with a population of orangutans habituated to humans), he was watching a female with an infant as they foraged in the trees. After half an hour of quiet observation, he noticed a mature male approaching her. The female, who was nursing and therefore could not have been in estrus, stopped foraging to watch the male with seeming casual interest.

He came on slowly, swinging from branch to branch, climbing with all four hands down the trunks of trees to reach her elevation just above the ground. She turned her back to him while bringing her infant up to her face to kiss and embrace. As she cuddled the infant, the male came around in front of her, put both arms around her and the infant, and hugged them for several minutes. Still holding the infant between them, she reached with one hand to touch the male's arm while turning her face up to his. The male kissed her squarely on the lips, a long, lingering, seemingly romantic kiss. Then they separated a few inches so she could look straight at him with what seemed to be a smile while the male looked down. Harding tried not to be anthropomorphic about this, but he believed the male looked bashful—there was no other way to describe it. Then the male turned and swung off into the forest. Was he an old friend of the female, a former or prospective future mate, or a brother? It is impossible to say, but no one can observe an intimate incident such as this and take at face value the assertion that male orangutans have no regard for infants and no interest in females except for sex.

Sons and Mothers

It has already been noted that food is the holy grail for females, as sex is for males. In any one species, therefore, males and females may be working at cross-purposes. Yet in primates especially, some species do have close male and female relationships, as described earlier. In species in which males remain within their natal group when they grow up but adolescent females disperse, mothers and their adult sons may remain close. This makes sense genetically; the mother may be able to improve the life of her son, which will benefit both their genetic inheritances. Here, we consider bonobos, chimpanzees, and muriquis.[1]

Bonobos

Bonobos do not behave at all like common chimpanzees, as was seen in chapter 4. For one thing, male bonobos are more giving: They can be aggressive at times, but they also share food, doing so more often with females than with males (Kano, 1992). Groups of females may gang up on males, but male groups never retaliate. Females are perfectly willing to have sex with males (other than their sons) even if they are not in estrus. Freely available sex keeps the males happy—they have no reason not to get along well with the other sex. Physical sex, therefore, is disconnected from reproduction and serves a role in communication and tension reduction, a healthy result for the community (Paoli, 2009).

The connection between a grown son and his mother, if there is one, remains constant. Both individuals have sex with others, but that does not diminish the son-mother tie. Bonobos are more sex-crazed than other primates, but mothers and sons do not think of each other in a sexual way. When researchers first saw male and female bonobo couples traveling together, they assumed that this was a mated pair. What else would it be? It never occurred to them that a son could be so strongly bonded with his mother that he was psychologically bound to her for life. Yet when adult male and adult female bonobos are found together, they are twice as likely to be sons and mothers than any other type of duo (Furuichi & Ihobe, 1994). In most species, it is a daughter that is closer to a mother, but in bonobos a daughter loses interest in her family as she grows up. As an adolescent, she increasingly spends more time at the edge of her group rather than near the others, until one day she leaves it entirely to join up with another. She starts a new life for herself.

By contrast, the adolescent male hangs around with his mother. He will be glad he does so because not only do they groom each other frequently, but if he tries to eat at a fruiting tree where there are other females, they will probably drive him away; females are dominant to males in this species and act accordingly. A son and his mother have to find food on their own (de Waal, 1997). When a male is fully adult, he still depends on his mother, who supports him in squabbles about dominance ranking, something males care about. (By contrast, females reduce group tension and enjoy life by having sex with each other.) If a male gets into a fight with another male, his mother will join in to make sure that, if possible, her son wins. The bond between a son and his mother is far stronger than that between an unrelated male and female whose brief liaison centers on sex.

Here is an example of tension in the world of bonobos, where the powerhouses are not the males but their mothers (de Waal, 1997): Kame was a senior mother living in Zaire with three sons who probably filled her with mixed emotions. She was a major player in their lives, but they were always hanging around wanting her help. Her oldest was Ibo, the alpha male, who was being challenged by Ten, the beta male whose mother is Sen. Kame was getting old, though, and could not fight the way she used to. By contrast, Sen is in fine health and, sensing Kame's weakening, she began to attack Kame's sons on behalf of her own son Ten. She even defeated Ibo himself once in one-on-one combat. But, the critical confrontation did not occur between Ibo and Ten, the two males, but between their mothers. Kame and Sen had a fierce battle that Sen won when she pinned Kame to the ground. The females often scrapped after this, but Kame had lost her drive. She never recovered from this defeat. Ten became the alpha male instead of Ibo, and Sen became the alpha female. If a female is high ranking, then the knowledge that she will help her son fight his battles means that he attains a higher rank in the male dominance hierarchy than he would had his mother been dead. Of course, this also means that he is in for a shock when his mother does die and his dominance ranking falls too. The sons of old Kame remained middle ranked for a few years until Kame's death; then they lost even that ranking.

Chimpanzees

Sometimes chimpanzee sons remain in their natal groups when they reach adulthood, and they too may remain close friends with their mothers. Such a mother was Flo, who was made famous by Jane Goodall and about whom Goodall wrote, "There can be no doubt that her status and personality were strong factors in the rise to power of her son Figan" (1986, p. 66). Even when Figan grew into middle age, Flo still worried about him. Once, when he was 23 years old, he hurt his arm while scrapping with another male. He screamed so loudly that Flo heard him from half a kilometer away. She ran towards him as fast as she could, although she herself was getting old. When she arrived at his side, he was still wailing, so she sat down to groom him. Gradually his cries tapered off as he calmed down. Earlier, Figan had attacked his mother several times at the banana feeding station where aggression was rampant, but she obviously had forgiven him. Their tight psychic bond lasted for their lifetime.

In the Taï National Forest of Ivory Coast, adult sons were also supported by their mothers. Christophe Boesch (2009) reported that high-ranking mothers had a new infant 2 years later if they had a son to raise rather than a daughter. By nursing longer, these young males were more likely than other male youngsters to reach adulthood and become high ranking with a good chance of having many offspring. Highly dominant Ella produced four sons in a row; the two oldest became alpha males of their group, together fathering seven offspring.

Muriquis

Another species in which sons remain attached to their mothers when they reach adulthood is the muriqui, or woolly spider monkey, of South America. Karen Strier (1992, p. 77) noticed this when she was teaching students to recognize individuals in the forest so that they could continue with the behavioral work she had started. The watchers were sitting together along a path early one morning, waiting for the monkeys to wake up and begin their day. As they could see animals scattered in the trees above them, Strier began quizzing the students on their identities. They named Diego correctly, then struggled with the name for the female who had swung over to join him on a branch. "Didi," they finally decided, which was the correct answer. By this time, Diego and Didi had exchanged a long, leisurely embrace and were sitting comfortably side by side. Although the students were not too excited by this association, Strier was, because Diego was Didi's son. She had known Diego when he was just a youngster, 9 years earlier, and was thrilled to see that he was now a robust adult who was still close friends with his mother.

RARITY OF FORCED COPULATIONS

In their book *Demonic Males: Apes and the Origins of Human Violence*, aggro-men Richard Wrangham and Dale Peterson (1996) wrote, "Even though rape is unusually common for apes generally, each of the five species shows a distinctive distribution for the behavior" (p. 142). Although this statement fits with the thesis of their book, it does not fit with the facts. Rape, hereafter called forced copulation because rape is a legal term, is not an evolutionary strategy. Forced copulation is very rare among nonhuman apes, with the occasional exception of orangutan females, which live alone and may be forced into sex by an adolescent male if there is no mature male nearby (Galdikas, 1985).[2] The females much prefer to mate with mature males, where the copulation is languorous, beginning with oral or manual manipulation of the partner's genitalia, and initiated by either participant. The couple progresses to full sex that they accomplish high in the forest canopy, the female hanging from a high branch while reclining against a lower one (Wrangham & Peterson, 1996). This enjoyable activity may continue for half an hour. Anne Maggioncalda and Robert Sapolsky (2002) stated that "close observations of orangutan rape show that it is very different from human rape: for example, researchers have never seen a male orangutan injure a female during copulation in an apparently intentional manner" (p. 60). Moreover, Harding (2011) reported affiliative (friendly) behaviors between adult male and female orangutans that call into question the inference that rape is frequent and males are violent to females.

What is the situation for the other apes? For bonobos, there is no forced copulation. Why would there be when the females are as interested in and as eager to have sex as the males are? Forced copulation is almost unknown in gorillas, which live in peaceful small groups with one dominant male to father the group's young (Robbins, 2009). Forced copulation does not occur at all among the 16 or so species of gibbons; they live in peaceful, monogamous relationships, and the females initiate copulation.

For chimpanzees, forced copulation is almost nonexistent, as evidenced by the legions of researchers who have observed chimpanzee behavior over 50 years. When females are in estrus, they mate with any number of males, as many as 14 in a day, but they are most reluctant to commit incest. When Fifi was in estrus, she willingly presented to and copulated with most of the males in her group, but she resisted mating with her brother, Figan (Goodall, 1986). On one occasion when he approached, she screamed and tried to jump from the tree she was on. He copulated with her as best he could while she hung from a branch. On another occasion, he trapped her among vegetation on the ground from which she could not break free. Goblin also forced himself on his sister, Gremlin. It is impossible to know if Figan and Goblin would have acted so aggressively against their sisters had they not suffered the frustration earlier of being provisioned and yet often denied bananas in the feeding program at Gombe; forced copulation has not been observed among chimpanzees of the Budongo Forest, where free food has never been offered to be fought over (Reynolds, 2005).

Baboon males are among the most aggressive primates, but even they do not force copulation. Barbara Smuts (1985) postulated why this is so (although she also noted that females can prevent it by the powerful act of simply sitting down): If a male X attacked a female in estrus as if to forcibly copulate with her, she would run away screaming. Other males that noticed her predicament might easily try to mate with her themselves, so X would likely lose her. With the same scenario, the female's screams might attract her male friend with whom she had been friendly for some time. The friend might attack X, which would also prevent him from mating. The female's screams from an attempted unwanted copulation would likely foster a mob attack on X by other females and their young. This would not only disrupt X's plan but would attract other males, which might join the mob and alert other females that X was a hateful individual with whom they should not mate in the future.

In summary, in nuclear families, the members interact closely to raise their young, as happens usually in human beings. In polygamous species, by contrast, females are busy raising their young while males have their own agendas. Even so, a variety of species, such as those considered in this chapter, have dyads in which males form close, nonsexual friendships with females. The males are not aggressive in these relationships but rather are supportive and empathetic. Males often have close male friends, too, the subject of the next chapter.

ENDNOTES

1. Predators may also have mothers and sons with tight bonds. This has been observed for coyotes and for leopards (Dagg, 2011).
2. In 2000, Randy Thornhill and Craig Palmer published a book entitled *A Natural History of Rape: Biological Bases of Sexual Coercion* which argues that for nonhuman animals and people alike, rape is an evolutionary strategy of males to increase their reproductive success. However, their book is full of misinformation, misogyny, and bias. For an analysis of its inadequacies, see chapter 9 of Dagg's 2005 book *"Love of Shopping" Is Not a Gene: Problems With Darwinian Psychology*.

Chapter 9

Male Friends

It would be great to be able to say that males that live together in a group always get along well together, but this would not be true. In evolutionary terms, males basically are out for themselves because each needs to produce a number of offspring that survive to reproduce themselves. For polygamous species, this usually means that males mate with as many females as possible, thus bringing them into competition with each other for breeding rights.

Copulations, however, are quick, and for most species, mating possibilities last for only a few days each year. There is ample opportunity for adult males to get along together at other times if they choose to, as most do; communities of primates leading tranquil lives are ubiquitous. Males may squabble, especially when there is a female in estrus nearby, but for primate species living in groups as they do for defense against predators, males need to be social (or at least nonconfrontational) most of the time; otherwise, their community would become dysfunctional. Cases of males attacking and killing other males within these communities are extremely rare. If controversy erupts, it is dealt with by threats or fights, after which calm returns so that community members can again go about their normal business of traveling, feeding, resting, grooming, and sleeping (Van Hooff & Van Schaik, 1994).

This chapter is about primate friendships, but "friend" is a human term. Because one can never know what a nonhuman animal is thinking, its motives, intentions, and feelings must be inferred. Wildlife ethologists (scientists who study behavior) often use terms that are intended to avoid anthropomorphisms. "Dyad," for example, means two individuals that are often together, without implying anything else about their relationship. However, as was seen in chapter 2, humans have the same glands, producing the same hormones that engender trust and impart a satisfied or contented feeling when they are in the company of friends. It would be absurd to think that humanlike feelings do not occur in nonhuman primates.

In studying friendships, though, one must be careful to ensure that male coalitions and mutual support are not based on mere convenience. In their research in China, where Tibetan macaques live in large multimale-multifemale groups, Carol Berman and her colleagues (2007) found that the alpha, beta, and gamma males supported each other not because they were good friends, but so they could corner the access to fertile females. On average, the alpha male accounted for about 64% of matings within his group, the beta male accounted for 21%, and the gamma male accounted for 8%. This left 7% of the matings for the males of lesser status. By sticking together, the three top males ensured that they mated most often with the females, and together they could prevent the lesser macaques from revolting against their dominance. They may or may not have been friends.

An example of a species whose males get along well as long as there are no females about is the blue monkey—a primate from the rain forests of western Kenya that is not actually blue, though it does have a blueish tinge to its face. The males often hang out together in easy companionship—with no tension, resting, perhaps grooming one another as the spirit moves them. When a female arrives, however, their relaxed mood vanishes. Patas monkey males behave in this same way. Further out in the relaxed-attitude scale is the muriqui of South America, discussed later, whose males also gather sociably in small groups. Should a lusty female join them to choose one with which to mate, the others watch with casual interest as the two copulate; maybe their turn will be next, or maybe not. Aggression in this species is almost unknown (Strier, 1992).

In this chapter, we look at six primate species that have evolved especially close friendships among the males. Pairs of males that are likely to be friends have grown up in the same group; those that are related, such as fathers and sons or brothers, are especially likely to form friendships. Spider monkeys represent small New World monkeys that are unaggressive by nature and whose males form subgroups that do not include females. Muriqui males are even more closely bonded. Bonnet monkey males are not especially friendly toward each other during their daily activities, but when it comes to sleeping, they crowd together. Among yellow baboons, two males may take a special liking to each other, forming in effect a bonded pair. Some gorillas live in all-male groups for years; other father-and-son pairs may share membership in a mixed herd that includes females and their young. In chimpanzees, humans' closest relatives along with bonobos, the males bond together for the purposes of travel, hunting, and boundary patrol of their territory.

The behavior of apes that never form male bonds is almost as informative as that of those that do. These include the many gibbon species and the orangutan, Southeast Asian apes that diverged from the African ape lineage about 13 million years ago. Gibbons are monogamous and territorial; because they live in bonded pairs that defend their territories against other males and females, males do not form coalitions. Coalitions are also impossible for orangutan males, which are solitary as adults except when meeting a female for mating. Therefore, male bonding is neither universal in apes, nor does it necessarily represent the ancestral condition. Human beings, chimpanzees, and bonobos are the most closely related of living hominids, but they share few inherited behaviors. The ancestors of human beings who moved from forests into a savannah habitat undoubtedly evolved their tendency toward and variable type of male bonding independently from those of other apes.

SPIDER MONKEYS

Until about the middle of the last century, some zoologists studied the behavior of a species by observing animals living in zoos. This is a lot easier and cheaper than trekking into foreign countries to spend years trying to find, identify, and describe what scores of active individuals are up to each day. Observations of captive individuals can provide useful insights into behavior, but the practice of many zoos was to place a male and a female or two or three in a cage and wait for young to be born that would attract more human visitors. Small species such as spider monkeys were popular because they were cheap to feed and cage compared to their larger and stronger relatives. Because most species do not live in couples in the wild, however, and are not fed and protected from predators, the behavioral findings of captives did not mean much. Such results for captive spider monkeys were especially misleading. In most primate species, males are somewhat distrustful of each other given their rivalry over mating rights. Yet for spider monkeys in the wild, it is the males, not the females, that form close bonds.

Male subgroups lead lives separate from and parallel to those of females. During the day, bands of a few males travel together, finding fruit and ensuring that their territory is safe from outside intruders. Should they encounter spider monkey invaders, each band, experienced in fighting, will likely drive them away. While they are resting, band males may embrace each other or sit in contact side by side. As was seen in chapter 5, senior bands of "brothers" that have grown up together since infanthood are immensely attractive to adolescent males, which may have been involved in "grappling" with older males and are anxious to join their subgroups.

However, spider monkey males are not perfect examples of peacefulness because, although they are devoted to each other, they are not so fond of females. A male may take the occasional swipe or cuff at a female, usually when two subgroups meet while foraging (Aureli & Schaffner, 2008). Because these attacks occur amid social excitement and the commotion of monkeys calling and leaping about in trees overhead, they make a vivid impression on human observers. Linda Fedigan and Margaret Baxter (1984), who studied the behavior of these monkeys in Tikal National Park, Guatemala, staring up into tree foliage week after week as furry individuals flashed by, reported only 10 attacks in 319 hours of watching the monkeys. The females that were chased and hit did not suffer any obvious injury; indeed, the attacks seemed to be stereotypical rather than an attempt to inflict harm, a sort of ritualized display.

Muriquis

There are two well-documented exceptions to the usual imperative of male rivalry about sex. Males of both lions and muriquis (woolly spider monkeys) are not rivals when it comes to copulating, but instead are polite, each waiting his turn to mate with a receptive female, hoping she will not get bored with the activity or approach some other aspiring male. Presumably the males do this because if they did not, there would be consequences. Male lions have sharp teeth and long claws; if two got into a fight over a female, one or both of them would be seriously hurt. Muriquis do not have such armaments, but they are hefty animals that live high up in tree canopies. If they began to tussle and one should fall, he would likely be killed. Common sense (alias evolutionary pressure) argues against aggressiveness toward their lusty rival males in these two species. Males profit by patience.

Muriquis live in Brazilian forests near the Atlantic Ocean, where Karen Strier (1992) studied their behavior for many years. She found these rare animals to be endlessly congenial. Males and females are about the same size, which in itself indicates that there is little aggression within the species. They hang out in fairly large groups, with the males usually congregating together when the group stops to rest or feed. They have known each other since birth (because it is the females rather than the males that emigrate on reaching puberty) and are therefore related to some extent with each other. Tolerance among this brotherhood means that dominance is unimportant. (Why would one worry about a hierarchy when all is peaceful and friendly?) They do not quarrel about mating rights and are quite willing to take turns copulating if a female is willing. Indeed, one theory to explain the close male friendships is that each wants to make sure he is around if one of his friends gets lucky with a female. Then he may get lucky, too. They are all related, so no matter whose sperm fertilizes a female's egg, all the males will have some of the same DNA passed on to the next generation.

Aggression among the group was almost nonexistent. During the 1,200 hours Strier spent watching these attractive monkeys during her first year, she saw only 31 interactions among group members that could be even remotely considered as aggressive. Nine of these were chases, usually when young females tried to join a new group. The other 22 were all of one type—one individual approached another and took his or her place. Strier saw no obvious body postures or facial expressions or calls that would indicate these were hostile "supplants." Moreover, they all involved different animals and so did not seem to be hierarchical in nature. But obviously, if these are considered aggressive encounters, then muriquis really are agreeable creatures.

Copulating as a muriqui sounds like fun if one is unafraid of heights. Their sessions last longer than in most species, 6 minutes on average but as many as 18 minutes on occasion (Strier, 1992). Lucky are the males that got to mate with Louise, perhaps the most lusty female because she had the most practice; she came into estrus and mated regularly without ever becoming pregnant. Strier noted that Louise liked to mate face to face, although she and her partner often had difficulty getting into a feasible position this way. At her most acrobatic and creative, Louise supported both herself and her partner from a branch with one arm while helping him to cling to her with the other.

Not all copulations end in ejaculation, but those that do produce a white ejaculate that sticks to the genitalia of both male and female. This white matter, which coagulates quickly, may form a copulatory plug in the female's vagina that helps to preclude a later mating—there are many variations in the sperm competition hypothesis. However, either one or both of the partners may decide to eat the plug as a good source of protein. Once a male tried to grab his own ejaculate from his partner, but she screamed and turned her back so that she could finish eating it herself.

Four male friends were pleased to connect with Cher, who had produced an infant 2 1/2 years earlier, weaned her daughter, and was again ready for sex. Before mating, a male inspects a female by tugging at her genitalia to stimulate a secretion that he sniffs or tastes directly with his mouth. If other males are near, they come to inspect her too, if she will allow this. Strier was also interested in Cher because she wanted to gather her feces to analyze for their ovarian hormone levels. She knew about Cher's activities because she had sat underneath her for nearly 6 hours, waiting for her to defecate. Cher had done so twice, but each time the dung had splattered or become mixed with that of other animals. (What a way for a researcher to spend a day!)

Finally, Cher was sprawled by herself on a low branch with Strier below, looking up with hope at her bottom. A group of males rested in a nearby tree. Clyde swung over to Cher, inspected her genitalia, then mounted her and ejaculated. Strier was amazed at how quickly this happened. Clyde left, to be replaced by Cutlip, who pulled out Clyde's ejaculate to chew, then mounted Cher himself. Cher took some ejaculate for herself while some fell to the ground, where Strier, delighted (scientifically, of course), collected it in a vial of alcohol. Cutlip then departed, leaving a plug in Cher's vagina that Preta pulled out and shared with Cher before mounting her himself. Finally, Scruff ambled up to remove the third plug and mate with Cher. There were no other males around, so his copulatory plug remained in place. To top off this sterling performance for Strier, Cher walked to the end of the branch, defecated a clean sample for her, then swung off to feed, leaving behind four sated males and an ecstatic researcher. All these males had copulated with Cher in one 11-minute period. The males remained best friends, but theoretically their copious semen was duking it out in Cher's reproductive tract (sperm competition) to determine whose sperm would sire her next infant.

Muriquis usually sleep on their haunches among high tree branches, several animals nestled so closely together that it is impossible to tell male from female or an adult from a youngster. Often, males have a group hug that sounds sentimental, and may be, but that can also be interpreted as males testing each others' strength (Strier, Dib, & Figueira, 2002).

BONNET MONKEYS

Like muriquis, male bonnet monkeys (or bonnet macaques) of India have a similar form of physical intimacy—a behavior called huddling, clasped-sleeping, or lumping. Bonnet males are not especially interested in grooming each other (although they often groom with females), and indeed, on occasion, males have serious fights in the wild (Silk, 1994, p. 288). But they do love to sleep in a bunch. In one study in India of a troop of 23 monkeys, males almost always slept closely with other males (and females with females); the only exception was the three biggest males, which slept alone (Koyama, 1973). Closeness defines the activity. At first, two males huddle together, clasped in each other's arms. Then other males arrive to attach themselves to the end of the train that soon spreads along a large branch. As many as 10 males, tightly pressed in a row, can slumber peacefully along one tree limb or occasionally on the ground. This closeness is apparently a response to possible predation. In the daytime, these huddles usually occur near tree trunks, but at night, when the animals are more vulnerable, they tend to sleep near the ends of branches, where predators such as leopards are too heavy to crawl out and reach them. This physical familiarity of bonnet monkeys, demonstrated in clasped-sleeping, undoubtedly helps to maintain cohesion within a troop.

Olive Baboons

Because they are terrestrial, olive baboons are more aggressive than most other monkeys, but even so, two males may team up to become close friends. Such a pair was Boz and Alexander, who were older than most of their fellows of the Eburru Cliffs troop living near Gilgil in Kenya (Smuts, 1985). Each morning, on waking, they searched each other out for a ritual greeting that involved approaching their friend with a rapid swinging gait, staring into his eyes, and often smacking his lips. Then one of them grasped the other's hips, played with his genitals, and/or mounted him. The other males sometimes performed these actions too, but they were less enthusiastic, with one participant often breaking them off before completion. Boz and Alexander not only greeted each other much more often than did other duos, but they were the only pair that took equal turns, showing complete symmetry in active and passive roles. The two friends cooperated occasionally in luring females in estrus away from prime males so they could copulate with them instead. They also protected each other from aggressive, more high-ranking males; when Boz, the elder, heard Alexander scream from 50 meters away on one occasion, he rushed to defend him, leaping onto the back of the assailant and driving him away.

Mountain Gorillas

Male mountain gorillas that are unable to attract any female friends wander about alone or join an all-male group. (Females are never alone; they are always in a group with at least one adult male that will not allow any other males to join it.) One such bachelor group led by the mature silverbacks Peanuts and Beatsme, made famous by Dian Fossey, comprised six unrelated males (Yamagiwa, 1987). These animals formed a cohesive whole, with the members even physically closer in general than in mixed-sex groups. Their activities were highly synchronized, including each day foraging, traveling, resting, playing, and making night nests near each other for sleeping. The members were more combative than the males in family groups, though, because there was jealousy about who mated with whom during homosexual behavior. The older males usually competed for and mounted the younger ones, and eight times in 3 months Peanuts and Beatsme fought each other. However, there were no major injuries, and the group carried on as an entity for at least 3 years; surely none lasts forever because such male groups are obviously a reproductive dead-end. However, they are useful for young animals which otherwise face the hazards of moving about alone as they mature.

Usually fathers and adult sons do not live together. The son leaves his natal family as an adolescent to join another group, so the issue of father-son rivalry for breeding rights with females rarely comes up. But sometimes it does. Beethoven and Icarus, both of whom have been described previously, lived in Group 5 as adults. Beethoven was the father, but a father growing old, so that he did not seem to worry if Icarus mated with Group 5 females. He had not liked it at first when Icarus was young, though. He would break up Pantsy and Icarus if they were copulating in order to mount Pantsy himself. But over the years, the two males became mellow. They had spent a lifetime together, traveling to find food every day, occasionally working together to defeat foreign gorillas that threatened the family. Once there was a violent fight with a neighboring group that left both father and son badly injured. They spent the next few weeks lying near each other, vocalizing gently together, recuperating from their wounds. Because young Icarus recovered first, he could at that point have challenged his broken father for the leadership of the family, but he did not do this, even though it took 6 months for Beethoven to heal completely.

After that, Beethoven depended more heavily on his son, especially when increasing tourism upset the animals. The pair learned to scatter groups of tourists and their armed guards with bluff charges. Icarus began to take over all the breeding duties with his half- and full sisters (Fossey, 1983). This was not ideal, but when his father died in 1985, he was free to hunt for other females that were not related to him.

Makale and Mutesi were brothers and close friends who lived with their family in the curiously named Bwindi Impenetrable National Park of southern Uganda. Thor Hanson (2008) was an American who came to know them while he was serving in the Peace Corps from 1993 to 1995. His job was to habituate wild gorillas to the presence of human beings so that the park could make money from tourists who were willing to pay a great deal of money to visit them. Makale and Mutesi's family comprised an alpha male, a female, and five younger males, which was too many males for the two brothers. Together they left the family to spend their days wandering in the park in search of females that might be persuaded to leave their current situation and team up with one of them.

Hanson had to leave Uganda before they managed to do this, but when he returned 12 years later, he was pleased to find that Makale had indeed been successful. He no longer associated with Mutesi, who was his older but less aggressive brother, but now had two female companions. Hanson followed his old comrade through the forest and noted from the group's night nests that there were now five gorillas in the family: Makale, two females, a juvenile, and one infant. Hanson hoped that Mutesi had been as fortunate in life as his brother.

Chimpanzees

Chimpanzee males are more sociable than females, which spend most of their time alone rearing their offspring. Females usually leave their natal group when they reach adolescence (thus preventing inbreeding), whereas the males usually stay, along with their male friends that had been their youthful playmates. Male adults compete vigorously for social dominance (Goodall, 1986, p. 162), but once they have sorted out their rankings in the dominance hierarchy, they are usually relaxed in each other's company, spending a great deal of time feeding, grooming, and traveling together. Chimpanzees have a fission-fusion society, with members joining up and leaving small subgroups at random, but at intervals they congregate in large gatherings where they meet and interact with a large number of individuals in the community to play, groom, display, and make a lot of noise. These gatherings can last a week or more, with chimpanzees coming and going as they please. They are the hub of chimpanzee life, rather like summer religious revival meetings in the early days of rural America, West Coast First Nations' potlatches, or the myriad other festivals people enjoy today. Male chimpanzees are more involved with these than females are, joining more gatherings on different days and staying for longer periods. When males left the gathering, it was often to seek a consortship with a female in estrus.

When they are not socializing, chimpanzee males regularly patrol the boundary of their territory in small male groups, sometimes accompanied by a female in estrus, to make sure alien chimpanzees have not crossed it (Goodall, 1986). Such patrols are easy to spot because the animals move along cautiously, making no vocal sounds and avoiding stepping on dry leaves that might rustle. They pause often to look around and listen. Sometimes they sit quietly in tall trees for an hour or more, gazing out beyond their territory. If they come across feces or discarded bits of food, they pick them up and smell them. If they see a recent nest in a tree, they climb up to inspect it before tearing it apart. All of their movements are controlled and tense. If they are startled by a sudden sound, they often reach out to touch or embrace each other. Returning from a patrol is a different story. With the tension released, the males call loudly, give drumming displays, hurl rocks, and chase each other about. Their exuberance resembles that of young men returning victorious from a football game.

Hunting was another activity that was undertaken mostly by male coalitions. In one episode described by Jane Goodall (1986), six males, including Goblin, Satan, Jomeo, Figan, and Sherry, noticed a baboon female and her young feeding in a palm tree. Goblin reached over to touch

Satan as they all stared up at her, their hair erect. Jomeo slowly climbed a neighboring palm tree until he was level with the baboons. He shook a branch, perhaps to make her run, but she stayed where she was and screeched instead, although no other baboons were within earshot to help her. Slowly, Figan and Sherry climbed the only two other trees within jumping distance from hers. The rest of them waited on the ground. When they were all in place, Jomeo leaped into her tree and the baboon jumped over to Figan's, where she was easily grabbed and her infant seized. The mother managed to run off screaming while the males tore apart and ate her youngster.

Although males in general team up with other males to hang out or to go on patrol or on hunts, these are not necessarily always the same males. Individual males may be more or less warm towards their friends not only over females in estrus, but for other reasons too. Jane Goodall described this for two of her chimpanzees, Faben and Figan, both sons of Flo. When they were young, these brothers were close, and although Faben was 6 years older, he seldom bossed Figan about despite his dominance. Then Faben contracted polio, losing the use of one arm. Figan then became dominant to Faben, even harassing him on one occasion by shaking him out of a tree, so their friendship cooled. Three years later, they became somewhat more friendly, grooming each other on occasion, traveling together, and sometimes displaying in unison at other males. Four years after that, they had finally become close allies, with Figan attaining the rank of alpha male of his group, in part with the help of Faben. One wet and cold evening, the two brothers even constructed an extra-large night nest where they slept cuddled together, something that had never before been observed (Peterson, 2006).

The games that chimpanzees play as they are growing up continue into adulthood. Jane Goodall (1986) saw this soon after she arrived in Gombe to begin her research. David Greybeard and Goliath, both about 25 years old and thus in their prime, were stretched out together in the shade when suddenly Goliath leaned over and tickled his friend's hand. David responded, and for 2 minutes they finger-wrestled, although otherwise lying quite still on their backs. Then David rolled over to dig his fingers into Goliath's groin. Goliath responded by tickling David's neck. Soon both were laughing. After 3 minutes, they roused themselves completely, with Goliath chasing David five times around a tree trunk. Then they settled down to groom each other for the next 22 minutes. (Goodall was methodical in timing all events.)

Later, Goodall observed lots of play behavior among males. Once, 11-year-old Atlas encouraged Humphrey, then 32, to play with him by piling leaves near him, then trying to stand on his head in the middle of the heap. Humphrey grabbed at him, then set off to run around a tree trunk with Atlas after him. Suddenly Humphrey stopped and kicked back, sending Atlas flying through the air behind him. Undaunted, Atlas rushed back for more fun. They chased each other, stopping now and then to wrestle and tickle, both laughing loudly. Several times both turned somersaults. Their play stopped when Humphrey moved away to investigate a noise nearby. Goodall followed him to find Fifi and her family sprawled out on the ground. Seven-year-old Freud came running up to his uncle Figan, then 25 years old, wanting to play, whereupon Figan grabbed his foot. The two of them roughhoused for long minutes, wrestling, tickling, turning somersaults, and chasing each other. Fifi was too lazy to get up, but she joined the fun now and then by tickling her brother Figan or Freud if they came close. The game ended when Figan loped seven times in a wide circle around bushes and two trees, followed closely by Freud, with Fifi grabbing at his heels as he passed her. Then, presumably tired, he walked off and Freud returned to his mother.

Chimpanzees males in their prime are aggressive if they are riled up, perhaps when a female in estrus is in their midst or when food (bananas) is provided that provokes fights. When they grow old, they may become close friends with another oldster that was once their rival. Jane Goodall in her book *The Chimpanzees of Gombe* described three of such friendships.

Hugo was the oldest, a large chimp and a leader who would give aggressive threats at times if there were food about, but who was also quick to reach out and reassure his victim afterwards that he meant no harm. He and Mike, an alpha male for many years, became faithful grooming partners when they grew old. They died in a pneumonia epidemic when they were both more than 40 years old.

Humphrey was another alpha male, born about 1946. Like Hugo, he could be very aggressive, but this may have been because of a painful abscess in one ear that often caused him to sit hunched down with his finger or thumb pushed against the ear. For 4 years after he lost his alpha status and before he died, he formed a close friendship with Figan, an erstwhile rival who was a few years younger. Figan outlived him by a year, during which he befriended two other old males, Evered and Jomeo.

Old Leakey tended to protect and support Mr. Worzle, who was some years older and probably his brother. Neither had an aggressive temperament, so Leakey remained a midranked male, and Mr. Worzle, who was very small, was low ranked. They had taken turns to go off with Ollie to mate, Leakey one month and Mr. Worzle the next. Mr. Worzle died of a wasting disease, followed the next year in 1970 by Leakey, who died of pneumonia.

Men

What does all of this mean for aggression among human males? Primatologists agree that humans' large brains evolved as the size of groups of their early ancestors were increasing, most likely in response to the need to remember, organize, and manipulate increasingly complex social relationships (Kudo & Dunbar, 2001). A bigger group size means there is more opportunity for intragroup squabbles and implies the need for social mechanisms to avoid fights, as was seen in chapter 5 on types of appeasement. Larger clans appeared millions of years ago, after humans' male and female ancestors had evolved to be fairly similar in body weight and canine tooth size, changes that indicate a shift toward equality. Here can be seen the evolution of nuclear families within a clan or tribe containing many families, with perhaps a differentiation in gender roles as males work together part of the day to make tools, hunt, and so on and then rejoin their families at night.

The idea of evolution-induced gender roles may discomfit some people, but the proof is in the girdles—that is, pelvic girdles in women and shoulder girdles in men. As humans' ancestors' brains grew, females' pelvises had to enlarge to accommodate larger skulls during childbirth. Concurrently, neoteny—the delayed maturity of children—increased, meaning that to ensure succession, males had to take increasingly active roles in child rearing for longer periods. Meanwhile, their shoulder girdles expanded, proving a need for upper body strength. Why did this happen? Theories for its cause include roughhousing, making tools out of rocks, butchering large carcasses, and—our own idea—carrying toddlers on their shoulders: The history of *Homo sapiens* and the predecessor *H. erectus* involved nuclear families crossing continents while carrying babies, young children, and simple implements.

The male coalitions that exist in workplaces, pubs, sports locker rooms, and battlefields may be analogous to the goal-oriented coalitions of chimpanzees, but they are not the same as close friendships. Within any given sports team, for example, one is likely to find small subsets of men who are friends who also go fishing together, barbeque at each others' houses, and are well known to each others' children. Friendships that are formed in childhood or in the teenage years often transcend time and the exigencies of life. Many male bonds are forged on a shared struggle after adolescence: The harder the struggle, the tighter the bond. Judging from references in literature and poetry, probably no male bond is stronger than that formed on the battlefield.

The uniquely male dichotomy of humans' recent forebears and of ourselves is this: A man needs his family and he needs his friends, yet they occupy different realms. Being away from one's female for part of the day implies the need for trust that other males will not romance her or injure her children. Trust is also necessary for cooperation in toolmaking, hunting, and innumerable other joint activities that are needed for survival. Chimpanzees (although not bonobos) have male coalitions, and it is possible that humans' ancestors at the time of divergence, about 5.4 million years ago, did too. Trust emphatically did not extend to strangers, who could be dangerous threats. In the ensuing eons, males' need for trust in their fellows evolved into the tight friend-bonding that can be seen today. A friend is the person one calls for help—whether to haul something heavy to the dump, influence a politician, or support a merger. He is the one standing beside a man at his wedding or shedding "unmanly" tears at his funeral.

Section III

Disposing of Fallacies

The next two chapters address fundamental fallacies that have contributed much to the biases regarding male aggression: infanticide by males and infanticide by females. The concluding chapter, chapter 12, argues that male aggression today is far more likely to be related to culture than to inheritance.

Chapter 10

The Fallacy of Male Infanticide

If male primates are perceived to be nasty and aggressive, as they are by aggro-men, then it does not seem too far-fetched to accuse many of them of infanticide. After all, in the animal world infants disappear all the time because of diseases, accidents, becoming lost, or failing to thrive from birth. Being small, almost always they vanish without anyone knowing why. However, very rarely, some males have been seen to kill infants, either because they are in a bad mood or, in theory, so that the infant's mother will come into estrus sooner and mate with the killer male. This chapter describes how a hypothesis can be so taken to extremes that some monkey and ape males are accused of infanticide even though there is no evidence that it has ever occurred in their species. Females kill about as many infants as do males, as shall be seen in chapter 11, but no one is much interested in them, particularly aggro-men who are keen to focus on male human ancestors as aggressive killers. This perspective is not limited to aggro-men; some feminists who have viewed men as "the enemy" are quite willing to believe in the aggressive archetype.

At the Kuala Selangor Nature Park on the coast of Malaysia, Harding paused at the Visitors' Center for a breakfast of pulut: steamed rice wrapped in banana leaves. Silvered leaf monkeys, also known as lutungs, were gathered in the tree above, their black fur with grey ends giving them their "silvered" appearance. A lone adult male served as a lookout for the dozen or so females, some with orange infants on their stomachs making a startling color contrast. They foraged for edible leaves, mothers often handing morsels to their young, all the while keeping an eye on Harding. When he threw a banana leaf into a nearby trash can, the closest ones stared at it with considerable interest, but they were too timid to leave their tree to retrieve it. A trio of the more terrestrial long-tailed macaques walked up without fear of him to squabble garrulously over other banana leaves. As he departed, Harding placed his last banana leaf on a branch near the leaf monkeys. A nearby female—he could tell her gender by the pale hairs on her inner thighs—quickly grabbed it and began licking off the bits of rice. The male came over to her as if to snatch it, but he did not. Harding planned to write a scientific paper about silvered leaf monkeys, so he was thrilled to meet them so soon in his visit.

Over the next weeks, Harding observed these animals in all kinds of situations. They were the quietest and least aggressive monkeys he had ever seen, foraging peacefully in troops numbering from a dozen to two score individuals. Because their food supply was abundant, they had no need to fight over it. For days on end, Harding saw nothing more combative than a gentle swat from a male, much as a human father, reading a newspaper, might shoo away a pestering toddler. The females practiced allomothering, in which an adult may care for and nurse any of the infants in the troop. Instead of aggression, silvered leaf monkeys rely on frequent sex, friendly gestures, and vocal interactions to maintain group cohesion about such decisions as where to forage, when to rest, and where to sleep at night.

When Harding was writing up his paper about the behavior of this idyllic species (Harding, 2010), he was startled to come across an article with the following statement about silvered leaf monkeys: "Infanticide results from an adult male coming into the group, killing the resident adult male, and killing the infants dependent on their mothers so that the adult females will begin to ovulate again" ("Silvered Leaf-Monkey," 2008).

He wondered how this could be—surely these gentle creatures were not into killing each other? It was impossible to believe this could be so. Yukimaru Sugiyama (1967) had suggested such a possibility for hanuman langurs in 1967, and Sarah Hrdy (1974), who had seen male langurs killing infants, had agreed with him. She had written up her findings and proposed the infanticide-by-males hypothesis: When a new male expelled the resident male of a troop, he would then kill the young troop infants so that the females would soon come again into estrus. Then he would mate with them so that all of the troop infants in the near future would have his (aggressive) genes. Nasty males like him would have an evolutionary advantage over more pacific males that did not do this. Soon, infanticide would be widespread within the species. The hypothesis was so popular that Hrdy had a paper published about it in the prestigious journal *Science* (Hrdy & Hrdy, 1976), and then a further paper and reviews on the topic (Hrdy, 1976, 1979, 1992).

Two other factors seemed odd for the silvered leaf monkeys. First, if infanticide were at all common, one would expect lower infant survival rates in the wild than in zoos, where potentially infanticidal males are kept apart from females with infants. But this is not the case. The survival rate of infants in the wild is 70%, barely less than that in captivity—70.9% (Shelmidine, Borries, & McCann, 2009; Wolf, 1984). Second, the species' biology is such that infanticide will not induce the infant's mother to cease lactating and therefore come into estrus. Because mothers nurse other infants as well as their own, a female may stop lactating before her infant is weaned, or she may continue to lactate (and hence not ovulate) even after her cub is weaned or killed.

Together, we discussed this topic of infanticide, which was of particular interest to Dagg, who had earlier written several papers about it for lions, as shall be seen in chapter 11. It is impossible to know for sure why some lions kill infants. There is obviously much more to the story than just the one hypothesis that several later researchers happily accepted as the reason for the few known killings (Dagg, 1998, 2000; Packer, 2000). For example, lion males may kill because they are stressed: Two filmmakers hoping to see cubs snuffed out for a documentary followed two lion groups, which had to put up with this harassment day after day for many weeks, until the couple (unfortunately) succeeded in their aim (Packer, 1994).

However, the infanticide-by-males hypothesis was so easy to understand and so new and exciting that dozens of researchers began to reanalyze the behavioral results of "their" species to find if they, too, could have been affected by bloodthirsty males. (With more papers published, academic authors increase their reputations as prolific researchers even if the papers include suspect information. With increased reputations come promotions, honors, and more money for more research.) The hypothesis implied that many males are inherently aggressive and infanticidal—that they are rewarded for killing infants other than their own by (a) getting more sex and (b) passing on more of their genes to the next generation. It took the nascent field of sociobiology by storm.

It even justified falsification. For example, Kevin Bales (1980) reported that for both yellow and anubis baboons, "the taking over of a troop by an outside male includes the killing of the infants of that troop" (p. 458). However, as David Stein protested in his 1984 book about these baboons, "I know of no basis for this claim either from the literature or my own observations" (1984b, p. 172). Craig Packer (1979) agreed that in 10 years of observing olive baboons at Gombe, he had never seen infanticide by immigrating males. (However, male chacma baboons have killed infants in Botswana [Cheney & Seyfarth, 2007].)

In the eyes of zoologists studying animal behavior, as the males of many species began to seem nasty, that nastiness bubbled over sometimes to include men. Of course, it also captured the public imagination. Here, it seemed, was an evolutionary explanation for various otherwise disagreeable behaviors in humans. Men are inherently aggressive, even murderous, and would use every means, no matter how heinous, to spread their genes were they not thwarted by laws and the threat of jail. It explained a lot: political fighting, war, adultery, violence against women. And, if men inherited a disposition to be murderous, they would almost certainly be selfish, greedy, chauvinistic, and misogynistic as well.

Many zoologists began to assume that the possibility of infanticide had affected the social organization of all sorts of species, not just langurs and lions. The need to protect their infants from infanticidal males—the infanticide avoidance hypothesis—was first mentioned speculatively by Takayoshi Kano in 1995 (de Waal, 1997). Soon, it was thought to have influenced every aspect of mating systems: whether a species is monogamous, polygamous, or promiscuous; how animals act when they are together and when they are apart; whether or not the female displays evidence of being in estrus (and therefore sexually receptive) by such things as having an enlarged vulval swelling; and whether infants and juveniles have a coloration that differs from that of the adults. Infanticide by males has become a "staple theorem of sociobiology … a litmus test upon which the validity of a sociobiological interpretation of behavior depends" (Sommer, 2000, p. 9); it may never have happened, but researchers could write that it would have occurred had not the society been organized to prevent it. How can one falsify this claim, as any scientist should be able to do if he or she accepts a new scientific fact? (From a contrarian point of view, should a female stick closely to one male in a multimale troop, as if to prevent infanticide, her behavior might actually advertise that this male is fathering her offspring, thus increasing the vulnerability of her offspring to infanticide from other males, which thus know they cannot be the father [Palombit, 2009].)

Harding decided to dig deeper into the issue of infanticide in the silvered leaf monkeys. For this species, only one instance of infanticide had been reported, by student Katherine Wolf in Kuala Selangor, Malaysia; she and a colleague had written it up (Wolf & Fleagle, 1977). Wolf had studied this species as a side product of her main research into other primates in the interior mountains of Malaysia, which were also home to leeches, rain, mud, and dense jungle. Far more pleasant than the mountains, the seaside park in Kuala Selangor (the very park Harding would later visit) has a small hill shaded by tall trees where three troops of leaf monkeys lived adjacent to a mangrove swamp that had been dyked to create walking trails. Other troops lived in adjacent territories up and down the coast, but oil palm plantations and other developments have since greatly decreased their habitat. Some unattached males also lived in Kuala Selagnor, keeping their distance from the male-female troops.

On breaks from her research on primates in the mountains, Wolf observed and made notes about these leaf monkeys. On one such occasion, after a 3-month absence, she noticed that a new male was in charge of one of the troops, the previous troop male was missing, and there were no infants. Could there have been an epidemic of disease or an accident to the male that left the group defenseless and at the mercy of predators? Could these even be animals from a different troop? But Hrdy had recently advanced the infanticide-by-males hypothesis, and Wolf was apparently eager to adopt it. She decided that there must have been infanticide in her troop too, judging from the evidence of the new male and the absent infants.

Unfortunately, her conjecture was soon treated as fact (e.g., Poirier, 1974). Infanticide by males is now assumed to be a standard behavior of silvered leaf monkeys as well as dusky leaf monkeys, although it has never been observed in either of these species. Both are assessed as "vulnerable to infanticide" despite the fact that the two aspects of the monkeys' biology and behavior described earlier would seem to preclude it (Van Noordwijk & van Schaik, 2000).

Could a "Scarlet I," denoting infanticide, also have been metaphorically applied to the foreheads of other leaf monkey relatives? Sure enough, in detailed studies of the behavior of 29 species of leaf monkeys in the wild, during which observers tallied thousands of hours of observations over many years, infanticide is almost never reported. In the two or three species in which it has been seen, it was extremely rare—far too rare to be explained by evolution and genetics.

Perhaps other families of primates have also been wrongly accused of systemic infanticide? We began to read the literature on gibbons, which, like human beings, are apes. Like leaf monkeys, they live in forests of Southeast Asia, and, like humans, they are more or less monogamous. In this family (Hylobatidae), none of the biological factors associated with infanticide by males are present (Geissmann & Anzenberger, 2009). For example, gibbons almost always live in families of an adult female, an adult male or two, and young of various ages. (In one genus, *Nomascus*, sometimes a male lives with two adult females; these seem to be unusual cases of small, fragmented populations living in poor habitat where the need to share scarce resources calls for creativity in mating [Fan et al., 2006; Zhou et al., 2008].) Pairs mate for life and both defend their territory against other gibbon families. Males do not try to take over other territories. Males have sex only (as far as is known) with their partners, and some help care for their young. Females usually have sex only with their partners, although cases of extrapair copulations have been documented. Males and females are about the same size, so males cannot dominate females.

As one would expect from this list, infanticide has never been seen in the wild in any of the 16 species of gibbons during countless hours of observations of each (Reichard, 1995). Yet, some of the most prominent proponents of the male infanticide/infanticide avoidance paradigms persist in stating that gibbon behavior and biology have evolved as a defense against infanticide (Chivers, 2001; Reichard, 2003b; Reichard & Sommer, 1997; van Schaik, 2000; van Schaik & Kappeler, 1997). Such dedication to obsolete theories without a shred of evidence belongs to mythology, not science.

The problem of dreaming up an avoidance hypothesis to explain something that has never occurred obviously has difficulties. Adrian Treves (1997) had the idea of correlating this hypothesis with the reality that newborn lutung monkeys often are born with odd colors. In all 18 lutung (*Trachypithecus sp*) species (Harding, 2010, summarizes the group's taxonomy), newborns have "flamboyant" (Treves, 1997, p. 47) orange fur and white skin on their hands, feet, and face. The skin color changes within days of birth to black or grey, as in adults, and the orange natal cloak changes to the adult color within 3 to 5 months (Bernstein, 1968). The conventional wisdom was that the contrasting infant coat color promotes alloparental care, for example by helping adults quickly find infants when danger threatens or when they fall from the trees, as occasionally happens during the excitement when two groups meet (Roonwal & Mohnot, 1977). Allomothering benefits a mother by relieving her temporarily from the time and energy of child care, and it benefits an adolescent helper[1] by giving her maternal experience prior to the birth of her own first infant (Fairbanks, 1990).

Evidently caught up in the new male infanticide fervor, Treves (1997) suggested that the contrasting pelage of infants might be an infanticidal avoidance strategy. A group would perhaps be especially likely to notice if an orange youngster were snatched from the group by a male imposter. This seemed an exciting possibility, so Treves compiled a list of the coat colors of infants and parents belonging to 138 species of primates in order to test three hypotheses: (a) that contrasting natal coats are a function of alloparental interest, (b) that they serve as infant defense, or (c) that they provide a paternity cloak.[2] He found that neonatal fur color contrasted with adult fur color in more than half of the species he examined, and these tended to be the species in which male infanticide had been reported. There was, however, no statistical association of contrasting coat color with allomothering. Trevis concluded that the only previously published hypothesis—that contrasting coats encouraged alloparental care—was not supported by the data, whereas infant defense was.

He was wrong. Compiling a list of the coat colors of 138 primate species seemed like a good start to the project, but a huge problem was that this data set was biased, especially in which species were omitted:

- In the list were only 49 species in which allomothering had been reported.
- Only three of the leaf monkey genus *Trachypithecus* were represented out of the 18 or so species, all of which allomother and for all of which infanticide is either unknown or vanishingly rare.
- No *Presbytis* species were included that have contrasting coats but do not practice allomothering.
- There were no *Nomascus* gibbon species included, and several species of *Hylobates* gibbons were also omitted, all of which have contrasting infant coat colors but none of which exhibit male infanticide (Harding, 2012). Moreover, allomothering is not possible in gibbons because almost all families have only one adult female.

Another problem was that Treves did not define his terms. For example, how contrasting did the infant and adult fur colors have to be? He also did not distinguish between merely caring for an infant by an aunt or sister and allomothering (the nursing of another female's infant), which is far rarer.

Subsequently, researchers have further analyzed the question of whether infant coat color is possibly correlated with infanticide. They found an association between allomothering and natal coat color, but no correlation between coat color and aggression (Rossa & Regan, 2000). Treves's theory was finally destroyed completely when the color vision of each species with contrasting natal coats was analyzed. The reflectance spectra from the fur and skin, the orange coloration that was conspicuous to humans, was instead cryptic amongst foliage to all males and many females of their own species (Sumner & Mollon, 2003).

The male infanticide/infant defense hypotheses have been so attractive that many biologists have overlooked the standard warnings against overgeneralization that are emphasized in every science class. Another example was that of Carola Borries and her colleagues, who, after documenting the genetic advantage to hanuman langur males of killing infants other than their own, then generalized the infanticide avoidance hypothesis to all anthropoid primates that live in multimale-multifemale or single male–multiple female groups (Borries, Launhardt, Epplen, Epplen & Winkler, et al., 1999). Clearly, this was a mistake because, as has been seen, it does not apply to either *Trachypithecus* or *Presbytis* leaf monkeys, which live in both types of groups.

Wait a minute: "All anthropoid primates" includes human beings, many of whom live in single male–multifemale families. In fact, Nina Jablonski (Jablonski & Chaplin, 1993) theorized that male infanticide was one of the forces that drove the evolution of bipedalism in humans' ancestors: Females stood erect and faced attacking males to protect their infants in a "threat-appeasement display." She overlooked the facts that (a) infanticide is vanishingly infrequent in humans and rare or absent in all other apes, and (b) humans, like gibbons, do not have the biology or mating system needed to postulate either infanticide or infanticide avoidance; human ancestors have not had them for at least the last 4 million years.

Other researchers have further extended Borries's theorem, stating, for example, that "infanticide amongst animals including humans reflects evolutionary adaptation…such as exploitation of the infant for cannibalistic purposes, or parental manipulation of progeny" (Sommer, 2000, p. 10). Can the suggestion that eating one's own children is an adaptive strategy in humans be taken seriously? Is the human mating system and reproductive/developmental biology really based on the threat of male infanticide?

Have the general ideas of male aggression and the need for females to protect against it unduly influenced thinking about the evolution of the human species? Unfortunately, obsolete and erroneous ideas from sociobiology have influenced thinking about modern human behavior among academics and among the public (Reese, 2009). Instead of obsessing about male aggression, it is time to speculate about how amiability among protohominids may have shaped human relationships.

The next chapter, which also deals with infanticide, shows that this behavior is not peculiar to males. Females are as likely to kill infants (but not their own) as are males, and they do so for a variety of reasons. This fact is little known, though; certainly, aggro-men are not keen to advertise that females are equally as aggressive as males toward the most vulnerable individuals. It must be emphasized, though, that infants are vulnerable to a multitude of factors, such as accident, disease, and predation; in many species, more than 50% of infants die each year soon after birth. Only because it has been so valorized and overreported by aggro-men and others does infanticide need to be considered here.

ENDNOTES

1. Allomothering may include nursing by adult females other than an infant's mother, but obviously not by adolescent females, who help in other ways.
2. "Paternity cloak" is a theoretical mating system in which the female behaves in such a way that her mates do not know, or her physiology does not reveal, who the father is. In theory, this either prevents infanticide because the father would not kill infants that might be his own or it encourages paternal support, for the same reason. Some humans behave this way, for example.

Chapter 11

Infanticide by Females —Too Common

This book showcases the dominance of sociability rather than aggression in primate societies, but as was seen in the last chapter, all is not sweetness and light. Infants in a few species may perhaps be murdered by a male because he feels ornery or because he hopes to increase his mating potential. Infants are also murdered by females, but of course for different reasons.

A few years after the hypothesis of infanticide by males took flight among zoologists studying animal behavior in the field, Dagg decided to reread the book *The Serengeti Lion*, by George Schaller (1972), who pioneered research on African lions. Male lions were by this time considered the poster species for the infanticide hypothesis, although few killings of cubs had ever been observed. Schaller's book was written before the hypothesis was postulated, so it would be interesting to see if the details of the lions' lives fit the theory.

Dagg was no stranger to lions, having begun her Africa wildlife studies in 1956. To her surprise, Schaller's text described four cub deaths caused by males, but five caused by females. In addition, females were often responsible for many more cub deaths: If there were a dearth of meat, for example, mothers often left their young to die of starvation. How could there be a rational hypothesis about males killing infants when more cubs were actually being finished off by females? In 1983, Dagg wrote a paper entitled "Lying about Lions" describing what she had found. (Today she would not use such an in-your-face title, but she was younger and feistier then and upset by what seemed to be scientific misinformation.) She sent the paper to several top journals, all of which refused to publish it. The infanticide-by-males hypothesis was so firmly entrenched by that time that no editor wanted to take a chance on a paper that questioned or refuted it.

Ten years later, three anthropologists at Washington University produced a paper taking a second look at the frenzy about infanticidal males among primates (Bartlett, Sussman, & Cheverud, 1993). For the hypothesis to be true, a male had to not only kill an infant, but then he had to wait around until its mother again came into estrus so he could mate with her, and over the long term, he had to leave more progeny than nonmurderous males. The scientists combed the extensive literature to see how many examples actually fit this model. They found 48 cases of infanticide that they discussed in detail, but in only eight was there strong evidence that the infanticidal males had sired the subsequent young of the females whose previous infants they had killed. At that time, there was no way of telling which male did father an infant from its DNA, so even the eight males may not have been the actual sire. These research findings have never been rebutted.

Fifteen years later, Dagg was asked to update and rewrite her lion paper by an editor who was equally as suspect as she at the excess use of the hypothesis. Her article appeared in the journal *American Anthropologist* in 1998, infuriating everyone who disagreed with her conclusions. But it is imperative in science to let research findings determine a hypothesis, not to formulate a pleasing hypothesis and then insist that even a lack of data (in this case, cubs believed to have been killed by males despite any evidence) supports it. Perhaps Thomas Huxley said it best: "The great tragedy of Science—the slaying of a beautiful hypothesis by an ugly fact" (1870, p. 167).

It is generally assumed that infanticide is a male thing (thanks to the work of aggro-men), but obviously from the example of lions, this is not so. Indeed, in one survey of rodents, members of three families (squirrels, common rats and mice, and cavies from South America), infanticide in the wild was reported to have been committed as often (in 14 species each) by females as by males, although for captive individuals, females killed more young than did males (Blumstein, 2000).

Infanticide by females occurs in many species and for a variety of reasons, as shall be seen. Young disappear from a community all the time —far more young are born than will grow to maturity—and seldom does a human observer know why. This chapter shows that if an infant is killed or even just disappears, it cannot be assumed that a male was the culprit. Here are examples of the many reasons why females kill infants that are not their own.

Tender Morsels

Newborn animals make a tasty dish, as lovers of squab and veal will attest. The most widely known example of infanticide—and indeed, cannibalism—by females is that of Passion and Pom, two chimpanzees whom Jane Goodall (1986) observed for many years at Gombe, Tanzania. Pom was born to Passion in 1965 and grew up despite Passion's poor mothering skills. When Pom was about 10, her mother started a killing spree among their group's young. When a mother produced a baby, Passion saw it not as a wonderful addition to their group, but as a tasty meal to be shared with Pom, who soon became her accomplice. Passion would stalk and struggle with a mother while Pom concentrated on grabbing her newborn baby. Usually they were victorious. They were not vindictive about their activity, just intent on success. On one occasion when the bereft mother,

Melissa, approached Passion and Pom while they were dining on her baby, Passion reached out to embrace her (seemingly sympathizing with her loss) and the two mothers briefly held hands before she and Pom carried on with their meal. During a 4-year period, Passion and Pom were seen to grab and kill three infants, and they may have killed as many as seven others that disappeared from the group during that time. (Adult male chimpanzees sometimes attacked and killed infants too, but they all belonged to mothers from other groups. The males did not copulate with the bereaved mothers so there is no evolutionary hypothesis to cover their behavior. Presumably it was just an anomaly.)

Black-tailed prairie dogs look cute when they are bustling about in busy communities, but their secret burrow life in South Dakota is a nightmare. In research by John Hoogland (1994), 84 different individuals during an 11-year period killed infants from more than a third of all litters, so it was not just a few psychotics on the loose that mothers had to worry about. Indeed, mothers were usually the killers. More than a fifth of the 65 infanticides were perpetrated by nursing females against their own nephews and nieces. Hoogland did not see most of these killings, but he recorded many. A marauding female would enter another mother's burrow and emerge 30 minutes later with blood on her face and a full stomach, obviously satisfied. Immediately, the bereft mother stopped all maternal behavior because her young were dead.

Why would a mother kill her close relatives? One reason is that food is scarce early in the spring when the young are born, yet little vegetation is available. The infant meat provides protein and rare minerals that are perhaps crucial for the killers' own stressful lactation periods. Perhaps also the infant's mother was not in top condition and might not have been able to raise her young anyway. While a mother is busy killing her relatives, this marauding female is obviously leaving her own young unattended, and it is surely retribution that in 9% of cases, another female was at the same time eating the killer's own infants.

Prairie dog killers have other benefits from infanticide because they live in colonies. Once they have killed the litter of a relative, this female will devote herself to the community as a whole rather than to her now-nonexistent family. She has time to sit above ground scanning the surroundings for hawks or coyotes, to patrol the colony against predators or possible prairie dog invaders from another colony, and to refurbish communal burrows. In addition, there will be more forage for the killer's young once they are old enough to eat vegetation. Because the females are seasonal breeders, giving birth only in the spring, the infanticide has nothing to do with males taking advantage of a bereaved mother.

Too Many Young in a Family

Perhaps no animals are more two-faced and competitive in nature than female grey meerkats. On the television program *Meerkat Manor*, they come across as sweet little creatures snuffling about on the ground to find insects and other morsels to eat (Clutton-Brock, 2007). Mother, sisters, daughters, and cousins spend their lives together, foraging, traveling, and sleeping each night in a communal burrow. When there are babies in the group, the females take turns babysitting them. Later, they rustle up food to give to the youngsters when they are old enough to follow the pack. They seem friendly enough among themselves, but most of the mothers are cold-blooded killers.

Tim Clutton-Brock, who carried out this study with the help of more than 50 volunteers, is an amazing man. Devoting his life to learning more about how animals behave, he has carried out major research projects on a variety of species with different social organizations, including insects, fish, birds, colobus monkeys, red deer, and rhinos. More recently, he has concentrated on meerkats, tiny carnivores (relatives of the mongoose) that are less than a foot tall when they are standing on their hind legs, which they often do. They live in large communities of up to 30 or so individuals in southwest Africa, where there is so little rain that the vegetation is sparse. It is easy to see and record what the animals are up to each day as they rest in the shade of bushes or scurry about hunting for insects, geckos, larvae, and spiders.

To understand social interactions between individuals in a community, researchers have to know who is who, and for the meerkats, this involved identifying 300 animals living in 14 different groups. The researchers implanted a tiny metal tag in the loose skin on the neck of each meerkat; the tag's number could be read by an electronic scanner, but the researchers also dabbed marks on different parts of the meerkats' bodies with hair dye so that they could identify each one quickly.

First, Clutton-Brock and his colleagues had to habituate the meerkats to their presence. When the meerkats became tame, they often climbed onto people for a better vantage point from which to keep watch for predators. In the book *Meerkat Manor* is a photograph of a female on duty standing on a man's head. Because food is scarce and therefore plays a vital role in meerkat society, the researchers taught the meerkats to step onto an electronic scale in exchange for a tiny portion of cooked egg or a drop of water. By weighing each individual in the morning, at noon, and in the evening, they could document how much he or she had eaten and therefore determine its foraging success.

Most females become pregnant at about the same time and each, needless to say, wants to raise her own offspring. However, there was never enough food available for all the young that were born to the community's females. Some would have to go.

Here is a representative example of how meerkats carry on. After the dominant female Vialli in the Whiskers group disappeared, probably taken by an eagle that lived too far away for clicks from Vialli's transmitter to be detected by antenna, the rest of the adult females raised Vialli's seven youngsters without her. Soon, after all the females had mated with males from other communities, Flower, Smithers, Mi Julie, and Ugly were pregnant. It was easy to tell this from their swelling abdomens as they stood or sat in the open on sentinel duty. Flower, the oldest and heaviest female, gave birth first. The researchers could tell this because she went into the burrow one night in her swollen condition and came out the next morning slim and trim, with blood staining her rump. She had taken over Vialli's position as the dominant female, but this did not save her pups: The other pregnant females killed all of them. Three weeks later, Mi Julie gave birth, only to have her pups killed by Smithers, who came up to the burrow entrance with blood on her face. Three days afterward, Smithers gave birth to have her pups killed by Ugly—it was always the pregnant females that were primed to be the murderers. Finally, 5 days later, Ugly gave birth to three pups when there were no more pregnant females to do them in. She and her pups were the lucky ones. All the females then helped to raise these three, including often suckling them.

Meerkat life is hard. Ugly's three pups did not survive to adulthood. The largest was taken early on by an eagle that swooped down from a thorn tree to grab it. The next was left behind when a rival group ambushed the Whiskers, who fled, leaving the cub unattended. Its head was bitten through by the dominant female of the attackers. Ugly's last pup was snatched by a goshawk. There was no new generation of meerkats that year because the young of all four females had met a violent end.

Too Many Young for Available Resources

Nestlings are easy to kill if, when they hatch, they are undeveloped and must be fed for a time by their parents. A neighbor can often sneak into their nest and finish them off. But precocial young that at hatching can walk, swim, and find their own food may also be at risk. Young ducks are especially vulnerable: They live on a small body of water, their fathers are long gone, and their mothers may be overwhelmed with huge families.

In a case one spring in central Japan, six families of spot-billed ducks lived on a small pond along with one gadwall duck family (Shimada, Kuwabara, Yamakoshi, & Shichi, 2002). This was a lot of birds (15 families per hectare), and this overcrowding, or perhaps even her own family of 12, triggered a spotted duck mother to go berserk. Over the course of 5 days, she attacked and killed 11 ducklings from the other families. She pursued a family group, grabbed a lagging duckling by the neck, and pecked its body or held it underwater while its mother unhelpfully quacked in alarm. If the victim broke free, she chased it to attack again, pecking it (sometimes with her own ducklings joining in) until it stopped moving. With this mass extermination, there was more food on the pond for the rest of the birds.

NONATTRACTION OF PREDATORS

This is another case of a mother duck killing infants, but it is given its own section because the researchers decided that this infanticide served a different purpose than the one described in the previous section (Prokop, Trnka, & Trnka, 2009). In this case, a common pochard female duck turned killer in Slovakia. She and her family of six lived on a pond along with two other pochard families on a much larger, less crowded pond than that of the spot-billed ducks. One day, the killer female attacked the sole duckling of another female, whose mother fled. The killer's assault was videotaped; for 4 minutes, she hit the infant 200 times on its head and neck. When it tried to escape, she chased it, lifted it repeatedly by its neck, held it underwater, and pecked its body. She rested only when the bedraggled body stopped moving entirely. This was surely overkill for one small duckling.

What was the killer's motive? Was she feeling ornery? Did she not like the mother of her prey? Did she want some exercise? Or was she showing her family who was the boss? The human observers/researchers speculated that because there was no crowding of ducks on the lake, the mother may have done what she did to reduce the attraction of predators such as hawks, herons, or weasels to the pond. The fewer tasty unrelated small birds there were swimming about, the better it was for her family.

Physical Crowding and Bad Luck

The lives of pinnipeds, which include walrus (family Odobenidae), seals (family Phocidae), and sea lions (eared seals: family Otariidae), are quite different than those of land mammals. They spend each year swimming about and hunting for fish to eat in the ocean or snoozing on land or pack ice. When the season arrives for the young to be born early each spring, the females clamber onto ice or a shore of rock or sand to give birth.

Many species, such as elephant seals and sea lions, gather in a rookery crowded with hundreds or even thousands of animals. These include not only the females and their newly born pups, but large males which have mapped out territories on the shore for themselves and a number of the females. There, they mate with the mothers that have newly given birth, their own young to be born exactly a year later. (Thank heavens women do not have to put up with this!) Jostling outside the territories are younger, nonterritorial males (Le Boeuf & Campagna, 1994).

Suitable land areas that can serve as rookeries are rare, so they tend to be crowded with pinnipeds. The largest and therefore oldest Steller sea lion females park themselves on flat rocks near the high tide mark to have their young where the surf spray cools them on a hot day. Such a location is highly sought after, so a female has to be careful that no other mother roughly competing for this spot injures her baby. She must ensure that the youngster does not wander far, where it could be bitten and killed, and that she is in a position to feed it whenever it is hungry so that it will grow quickly and be less susceptible to attack; true seal females (phocids) may fast while they nurse their young, in which case the lactation period is by necessity short—only 4 days in the hooded seal. Eared seal females nurse for months in the rookery, swimming out for fish to sustain themselves at intervals, before the youngster is ready to begin its own life at sea.

Often, seal mothers and their pups become separated on land in the crush of bodies, in which case the young usually die. Some of these lost pups try to steal milk from another female, but to do so jeopardizes the future of this female's own pup, so that more than half of such "orphans" are killed by mothers, either directly or by dying from their injuries. The "best" mothers are those which reserve their milk for their own young and so are the ones most likely to kill stray pups. Male seals can be lethal to pups too, sometimes crushing them as they pursue females with which to mate, biting them or even trying to mate with them. But this is manslaughter, not murder.

Crowding causes the death of many young pinnipeds even after they have left the birthing area. A walrus mother and her pup swim with their herd from one feeding area to another, at each stopping place having to haul themselves onshore among and over scores of other resting walrus. Many young and sometimes a mother, too, are jabbed by the tusks of an adult male or female that is annoyed at being disturbed, a thrust that can kill either a baby or a mother. One large walrus herd, on leaving a rookery where it had been ensconced, left behind 48 dead youngsters.

What does all of this mean from an evolutionary context? Males do not worry about killing infants because they are likely not their own; they were conceived a year earlier, probably by another male. Good mothers protect their own young at any cost, providing only it with milk; if other pups die, this need not concern them. Young mothers with little experience are the least able to protect their young, but they learn with time. Thus, the killing of infants has nothing to do with the famous hypothesis of male infanticide. The females do not come into estrus once their infants are killed—a male would have to wait a year for that. Also, pinnipeds do not cannibalize them. Pinniped young born in a rookery are killed because of bad luck—bad luck because of crowding, poor weather, aggressive neighbors, or nearby males fighting and courting.

RIVALRY

Infanticide in animals is rarely seen by people—infants are small, birth areas are sheltered, and a little life can be snuffed out in seconds. Many researchers have reported infants found dead, but almost always the perpetrators are unknown. (If the body is found, the culprit was likely not a predator.) Even so, there may be suspects, and these are at least as often female as male.

For example, there is the guira cuckoo, which is common in Brazil, a number of them flying about together and parking themselves on fence posts or scrabbling about on lawns. They are easily recognized as the birds with bad hair—males and females are identical, with spiky feathers sticking up in all directions on their heads, giving each a deranged look. Maybe they are crazed by the mayhem that is going on back home.

Guira cuckoos nest in bunches rather than in pairs, one group of as many as 13 birds joining forces. The females mate with the males to lay perhaps 15 or 20 eggs in the communal nest, each large egg weighing about 16% of a female's weight and so representing a considerable expenditure of her energy. Researchers Regina Macedo and her colleague C. Bianchi (1997) were intrigued by this group arrangement because there would presumably be rivalry about who mated with whom and whose hatchlings wolfed down food from which parents. Most curious, however, was that eggs and nestlings were routinely tossed out of the nest to the ground to their unfortunate demise. The researchers swept up sorry remains under the nests each morning.

Macedo and Bianchi decided to monitor a number of groups to find out what was going on. If the young of any one nest cup were all tossed out of it, leaving nothing of the next generation, the group could renest, which many of them did at least once, giving 98 breeding attempts. The researchers focused on the nine nests that lost all their nestlings, 48 in total, some dead in the nest cup and others lying mutilated on the ground below.

How could this be? There was enough food brought from adults to the nest, so the nestlings could not have starved to death; they died within a few days of hatching, so they did not yet need much food, anyway. Predation was not suspected. Nestlings from three other nests had been attacked by predators, but this calamity was easily recognized because the invaders had left behind broken eggs, crushed leaves, and scattered twigs that were not present at the nine nests.

Macedo and Bianchi concluded that the nestlings had been killed by members of the group—by infanticide. Were they killed by males or females? It was impossible to say. Members of both genders had reason if they had had no part in the production of the nestlings. Perhaps a male had not mated with a female that laid an egg. Would he be aware of this? A female would know that she had not laid an egg, so perhaps females were the more culpable. Either way, by losing all the nestlings, the group would then probably try again, and the culprit(s) would have another chance to be genetically involved.

DISPLAYING SEDUCTIVE DOMINANCE

Female house sparrows also kill infants. House sparrows are widespread, so presumably anyone could have carried out the research that José Veiga did (1990, 2004). However, few people would have the patience to set up 80 nest boxes and, for 3 years, dedicate more than 500 hours to observing the reproductive life of 100 pairs of sparrows. Veiga did this on a 1-hectare plot in Spain where he caught sparrows in a mist net and fixed colored bands on their legs so that he could identify individuals. Most bird couples reared two broods per breeding season, from April to August, both male and female bringing food to the nestlings (although the males were less industrious about this than the females). A few males mated "bigamously" with second females, but they rarely helped feed this second female's young. Veiga monitored each nest four to six times a day during hatching and once a day during the laying and nestling periods.

Veiga seldom saw infanticide with his own eyes, although he did once witness a female entering the nest of a monogamous pair and attacking the 10 to 11 day-old nestlings within it. Some hours later, their corpses were strewn on the ground beneath the nest. Five days after that, the intruding female started laying a new clutch of eggs in this same nest, attended by the male of the monogamous pair. Her aggression was vindicated.

For other cases of infanticide, Veiga had to work out which adults were present or absent and which dead nestlings had unambiguous peck marks on their bodies. He calculated that in all, 66 eggs or nestlings were destroyed by males and females that wanted to be raising young of their own. Killer females destroyed clutches in order to start their own brood within the same nest box and with the same territorial male. Those females that lived in a monogamous relationship were more likely to have their clutches destroyed by females than were the females that had no (bigamous) male in attendance. This choice was wise, because it came with a male to mate with and to help raise the second brood, as well as a nest box in which to do it.

But Veiga was not content with these findings. He researched infanticide among his sparrows for many more years, publishing an update in 2004. He now felt confident that females carried out more infanticides than did males and that they did so regularly. These killer females were older and more experienced than younger birds, yet the reproductive success of infanticidal females did not differ from that of noninfanticidal ones. What was the point of the killings, then? They allowed the killer to lay her eggs earlier than she would have done otherwise, but not early enough to allow time to raise an extra brood that season. Veiga concluded that by destroying the young of unrelated house sparrows, the killer female displays a dominance over other females that will make her attractive to mating males in the future.

Triggering Dispersal

Infanticide can trigger the emigration of females from their group, presumably to prevent interbreeding. This is true for sifakas, which live in Madagascar, where they are acknowledged as champion jumpers. They do not jump high but far, leaping as much as 30 feet from one bound to the next. Sifaka groups are small, seldom more than eight, because each group usually has only one breeding female (rarely two) given their harsh environment and because no adult female or male will mate with a relative. There is no inbreeding at all—known for sure because each animal in the four groups that were studied had its blood and DNA collected and analyzed (Morelli, King, Pochron, & Wright, 2009). Sifakas are very clear about this; there is no fooling around. If an adult lives in a group with only opposite sex relatives, as it may choose to do for years, then it does not reproduce. Apparently, an individual recognizes its relatives by its keen sense of smell, and they remain just friends, not sex partners. This changes if a new mating partner enters the troop.

As a result, emigration is not uncommon, occurring equally but one at a time for males and females from both natal and breeding groups. When a male or female enters a new group, its first task is to kill the infant(s) already in it. For males, this may be so that a female will come into estrus soon and mate with them, as discussed in chapter 10. For the females, it has another function. Researchers found that immediately after a new female arrived, the infant of the group's breeding female disappeared, presumably killed. Right after that, its mother left the group, her position to be filled by the killer immigrant female. On leaving the group, a female sifaka disperses to a new one; it may be nearby, but if a close one is composed of her relatives, it may be one that is some distance away. This more or less forced emigration for females is costly; a female is more likely to be caught by a predator because she does not know the foraging area of her new group or the location of food and water. However, dispersal can also be beneficial, spreading genetic inheritance, perhaps decreasing predator pressure, and perhaps encouraging the exploitation of a new territory of higher quality.

Male Infants Not Wanted

Spider monkeys live in groups that include a few males and a few more females. There is some evidence that adult females sometimes kill the male young of their compatriots (Vick, 2008), which would serve to reduce the number of males that reach adulthood. Adult males may also kill young males in their own group, especially if reproductive competition is high.

Not Enough Food for All

It is not known why female pride lions attack and kill cubs. That they do is known because George Schaller found their tracks leading to the small corpses. Perhaps the cubs were born to nomadic lions which the pride females disliked? Perhaps the pride mothers were in a bad mood because they had not eaten recently? A lack of meat does cause them to abandon their own offspring. When there were few prey animals in the area, lionesses ate what meat was available, leaving nothing for their young. Schaller (1972) wrote, "It was depressing to see a starving cub totter to its mother, each rib sharply outlined beneath its unkempt hide, and receive a vicious cuff instead of a bite to eat" (p. 100). If a lioness herself had enough to eat, she did not necessarily worry about her cubs that went hungry. Schaller concluded that the many cubs that disappeared from his study died from maternal neglect. They were simply abandoned, perhaps in a thicket, and their mothers returned to their social lives with their mothers, sisters, nieces, and cousins. The evolutionary theory is that it makes sense for a mother to desert her ailing young that might not survive anyway; she can then keep up her own health so that when good times return along with abundant prey, she can produce more cubs that will thrive.

Indigenous groups of human beings will also kill infants if this is considered a better alternative than allowing a newborn to survive. Among the !Kung tribes in Botswana, women usually had a baby every 4 years.

If one was born too close in time to the next oldest brother or sister, the mother might kill it so that her milk could continue to strengthen the toddler she already had (Shostak, 1981). Infanticide was also practiced by the Yanomami when food was limited (Tierney, 2000).

In conclusion, although many people assume that infanticide among nonhuman animals is a male trait that enhances their mating success, actually such killing is as common for females as for males, and it is done for a wide variety of reasons. It is by no means necessarily a sign of male aggression.

Chapter 12

Aggression as Culture

Why Are Men Aggressive Today?

The theme of this book is that humans' early male ancestors were seldom aggressive but rather lived in nuclear families in peaceful groups; groups survive, after all, only where cooperation is essential and aggression is minimal or absent. The preceding chapters show that humans evolved over millions of years living relatively peacefully, pair-bonded in nuclear families; that overt aggression only appears in the archaeological record many tens of thousands of years after humans became "fully behaviorally modern" and is more likely cultural than hereditary; that faulty reporting and interpretation of biology were the bases of the pervasive notion that humans and many other primates are inherently aggressive; and that cooperative and affiliative behaviors more accurately characterize most primates' daily lives, including humans'.

Why does our contention that protohominoids before the evolution of *Homo sapiens* were largely peaceful seem so unusual? It is because books and movies about history are full of wars and massacres carried out by men which have come to seem natural. Ever since Jane Goodall reported in 1974 that a few male chimpanzees killed other chimpanzees, this well-publicized "war" has given added credence to humans' assumed belligerent primate nature. Dozens of books continue to valorize male aggression today and in humans' past.

Evolution has given human beings the intellect—meaning the will and capability—to get along together in peaceable societies, but these abilities were honed over millions of years for a specific set of conditions: nuclear families in loose aggregations of perhaps 150 individuals ranged widely in scattered populations in woodlands and on savannahs. In this setting, cooperation on several societal scales—family, clan, and community—was essential for survival. By about 20,000 years ago, the blink of an eye in evolutionary terms, humans' intellect and dexterous hands provided the ability to fashion highly effective tools for hunting, gathering, fishing, and building abodes that extended longevity and shortened interbirth intervals. Populations grew and expanded. As long as there was vacant land for the less feisty or less well equipped to migrate to, all was well. But eventually, because there was too little land for all the hominids, Neanderthals and many other species were wiped out.

The basic difference between human beings and other animals is that animals' lives and behavior are shaped entirely by evolutionary principles and humans' are not. For nonhuman animals, evolution shapes first the lives of females; they must have the food and protection that enables them to produce and raise the next generation. The lives of males align themselves with this principle, but individuals may also compete with each other to father the next generation. These principles have successfully produced and governed millions of thriving species; evolution is a miraculous process whereby each species in the universe is adapted to its environment—except for human beings.

Humankind has a history of slaughter, strife, and cruelty, but these date back less than 12,000 years (Fry, 2006). We argue that the evolution of large brains and high intelligence has enabled *Homo sapiens* over time to

- realize that an act of aggression can bring an individual or a group an advantage they would otherwise not have ("might is right"), even though it goes against evolutionary principles.
- wipe out around the world large species of mammals as the human species expanded its range after emerging from Africa 50,000 years ago (Flannery, 2010). Mammoths, the last large species to survive, became extinct about 12,800 years ago.
- live in complex societies of increasing group size that foster aggression because of their density. In small groups where everyone knows each other, it is impossible for any one individual to insist he is superior if he is not. One individual will not be allowed to harm the group. All individuals are more or less equal. When groups become so large that all members are not known to each other, equality cannot be assured. The result of too many people living too close together has often been violence, a consequence for which evolution has not yet provided an effective solution.

About 10,000 years ago, new technologies and some fortuitous mutations gave humans' forebears the ability to develop and store food in one place. This permitted or demanded an increase in group size far beyond "community." Humans' natural ability and inclination to defuse aggression became less valued as chieftains began to recruit warriors to fight other groups for resources that were still scarce and sparsely distributed.

The possibility of aggression has always been present in the human lineage, but its disruptive presence until recently has been muted. Despite the equal size of male and female canine teeth in humans' ancestors (along with other evidence), indicating pair bonding and a lack of male-male combat, the size difference between men and women suggests that because males are on average larger than females, this must have had some survival advantage. Perhaps bigger males were important to a group in survival situations of famine or emigration. Perhaps females preferred larger men[1] for protection against predators or even other hominids; for hundreds of millennia, humans' forebear *Homo habilis* shared life in Africa with at least three *Australopithecus* species, and later on, *Homo erectus* lived alongside at least four other *Homo* species (see figure 2 in chapter 2).

MODERN MAN

Although humans' male ancestors were not combative, aggression by men today is widespread (though not universal) in business, where competition is *de rigueur*, and in many team sports, where hitting and hurting is widely acclaimed. That widespread aggression today is cultural rather than genetic is evidenced by at least four scenarios.

Some societies are aggressive, such as those from Papua New Guinea, where men frequently injure their wives (Dubois, 2010), and others are far more peaceful, such as Bushmen of Botswana (Shostak, 1981). Some nations were warlike at one time, such as Vikings from Norway, but peaceful later on.

Societies can be brought to a frenzy of violence by brainwashing from their leaders. Japan has always had a cohesive island culture that considered itself to have achieved a high level of culture in terms of art, architecture, and interpersonal relations, relative to other societies. Before World War II, urged on by their government, Japanese soldiers were guilty of horrendous massacres of Chinese people. During an attack on Nanjing, more than 300,000 men and women were brutally killed and up to 80,000 women were raped (Chang, 1997). As another example, although Hutu and Tutsi peoples had lived together relatively peacefully in Rwanda for many years, by organizing a media campaign against the Tutsis (who were often referred to as cockroaches), Hutu men were roused up in 1994 to kill 800,000 of them (Dallaire, 2003).

Soldiers in a war sometimes suffer from post-traumatic stress disorder (PTSD) because of the atrocities they are forced to see and commit (Bradshaw, 2010; Bradshaw & Schore, 2007). PTSD affects the brain in ways that are measurable, causing neurological dysfunction and even lifelong consequences that often lead to aggression. Typically, a young man is traumatized in battle, has not had a father figure for a large part of his adolescence and early manhood, and has as a male role model his military superior, who has no incentives to instill community-positive values. His emotional growth is so stunted that he is easily spurred to violence. As is well known to sociologists, violence can be passed down repeatedly to successive generations. PTSD symptoms can even occur in boys not because of war, but through being orphaned at an early age. Recent research linking PTSD-like symptoms to being orphaned has been documented in Kabul (Catani, Schauer, Elbert, Missmahl, Bette & Neuner, 2009), Kurdish Iraq (Ahmad & Mohamad, 1996), Rwanda (Schaal & Elbert, 2006), South Africa (Cluver & Gardner, 2006), and many other countries (Cluver & Gardner, 2007). Nations in sub-Saharan Africa have about 15 million children under the age of 18 who have lost one or both parents to the HIV/AIDS pandemic, and these represent only about 30% of all orphans in that region (United Nations Children's Fund & UNAIDS Secretariat, 2006). PTSD has even been documented in elephants (Bradshaw, Schore, Brown, Poole, & Moss, 2005). Young males that had witnessed their mothers and other members of their herd being slaughtered grew up to be rogue animals. They unpredictably attacked other members of their own species as well as humans and rhinoceros in ways that elephants leading normal lives never do. That this happens to animals as well as people proves not only that PTSD has a common neurological basis but that it and the ensuing violence are not hereditary.)

The abusive treatment of women by many men today (and of men by some women) is significant evidence that such activity is cultural rather than inherited. In other species, males do not kill females, and certainly not females carrying their own fetuses. Yet this occurs in some human couples. As people thousands of years ago formed communities of increasing size, men, because they are larger and stronger than women, were able to devise and promulgate cultural norms that valorized their gender at the expense of women. Preeminently, men founded religions that have controlled all aspects of women's lives—their inferior education and status, their leisure (if any were allowed), their financial worth, and their independence—to men's advantage. Gender inequities linger to this day in all human cultures.

One way in which men harm and humiliate women is by rape. As was seen in chapter 8, rape almost never occurs as a possible evolutionary trait and then does so in only a few species. In an American sample, 20% of college women reported having been raped (Brener, McMahon, Warren & Douglas, 1999). In a South African sample, 28% of men admitted they had raped someone at least once in their lives (York, 2011).

Widespread rape by combatants has been documented in Bangladesh, Burma, Columbia, Somalia, Burundi, Sierra Leone, Rwanda, Democratic Republic of Congo (a country that gave researchers a new acronym—REV, for "rape with extreme violence" [Mukwege & Nangini, 2009]), Liberia, Sudan, and Uganda. In the most extreme cases, 15% of the female population has been raped (Anema, Joffres, Mills, & Spiegel, 2008; *PloS Medicine* Editors, 2009).

Equally horrific are the cultural "wife beatings" and "honor killings" of young women in the Islamic world for such "crimes" as having premarital sex, flirting, or failing to serve a meal on time. Countries reporting "honor killings" include Iran, Turkey, Afghanistan, Iraq, Saudi Arabia, Egypt, Palestine, Jordan, Bangladesh, Algeria, Ecuador, Morocco, India, Israel, Ethiopia, Somalia, Uganda, the Balkans, Sweden, Holland, Germany, Italy, and Yemen (Ali, 2008). While Harding was working in Jordan from 2002 through 2006, about two dozen women were reported killed each year in the name of family honor, but the actual number was far greater because such deaths were often officially categorized as "suicide" or "accident" even though the Jordanian constitution grants leniency for "honor" crimes (Ali, 2008). In India, "dowry deaths" refer to the more than 5,000 brides who die annually because their husbands' families consider their dowries insufficient (Mayell, 2002, citing UNICEF data).

That many men today are violent cannot be contested. But, their aggression is largely a product of modern culture—of male-dominated religions, the density of urban populations, the pervasiveness of poverty, societal destabilization by war and economic crises including drought, lack of employment opportunities, and the valorization of male aggression in the media. It is not correlated with humans' primate ancestors, who lived simple lives in small groups, or with the behavior of chimpanzees and bonobos, which have their own hereditary lineages. Because human violence is cultural rather than genetic, there is the possibility of a more harmonious world in the future. Humans have evolved throughout millions of years to be essentially peaceful. Although each person retains the ancestral capability for aggression, humans' hormones, genes, and cultures have given them the means to subdue it and channel it to community-positive forms. While human violence is cultural, it can also have a biological basis in the effects of psychological trauma, such as witnessing brutality or being orphaned. These are environmental, not hereditary influences. The statistics cited earlier indicating the variation in different forms and rates of violence among cultures show that it is possible for societies and their legal institutions to learn to minimize aggression and avoid violence.

It is our hope that the ideas in this book, and the scholarly research behind them, will influence politicians and the business leaders who support them. They should realize, for example, that raising an army dooms a segment of their own population to domestic abuse—not only their soldiers, but the children who are left behind. They should realize that international policies that increase poverty in their trading partners' countries threaten their own welfare because social dysfunction increases violence at all levels and leads, ultimately, to failed states. They should realize that the security, prosperity, and rule of law that is good for their own families, neighborhoods, and communities is good for all societies.

ENDNOTES

1. Female preference in mates is termed "sexual selection" rather than "natural selection" when it encourages features in males that have no obvious survival value for the males, the standard example being the outrageous colors and plumage ornaments of birds. First proposed by Charles Darwin in 1871 to explain sexual dimorphism in humans, it is now a widely accepted principle throughout the animal world.

Scientific Names of Animals Discussed

Baboon, chacma: *Papio hamadryas ursinus* (Palombit, 2009)

Baboon, gelada: *Theropithecus gelada*

Baboon, Guinea: *Papio papio*

Baboon, hamadryas: *Papio hamadryas*

Baboon, olive: *Papio cynocephalus* (Smuts, 1985), *Papio anubis* (Strum, 1987), or *Papio hamadryas anubis* (Palombit, 2009)

Baboon, yellow: *Papio cynocephalus* (Stein, 1984a) or *Papio hamadryas cynocephalus* (Palombit, 2009)

Bonobo (pygmy chimpanzee): *Pan paniscus*

Chimpanzee (common): *Pan troglodytes*

Cuckoo, guira: *Guira guira*

Duck, spot-billed: *Anas poecilorhyncha*

Geese, Canada: *Branta canadensis*

Gibbon, black *crested*: *Nomascus concolor*

Gibbon, white-handed: *Hylobates lar*

Gorilla, mountain: *Gorilla gorilla beringei* (Fossey, 1983) or *Gorilla beringei beringei* (Robbins, 2009)

Hominid: a member of the *Hominidae*, the human family

Honimoid: a member of the superfamily *Hominoidea*, which includes the gibbon family, *Hylobatidae*, and the human family, *hominidae*.

Langur, hanuman: *Semnopithecus entellus*

Langur, Nilgiri: *Semnopithecus johnii* (formerly *Trachypithecus johnii* or *Presbytis johnii*)

Lemur, fork-tailed: *Phaner furcifer*

Lion: *Panthera leo*

Lutung (leaf monkey): *Trachypithecus* sp (see monkey, leaf; langur)

Macaque, Barbary: *Macaca sylvana*

Macaque, bonnet: *Macaca radiata*

Macaque, Japanese: *Macaca fuscata*

Macaque, rhesus: *Macaca mulatta*

Macaque, stumptail: *Macaca arctoides*

Macaque, Tibetan: *Macaca thibetana*

Meerkat: *Suricata suricatta*

Monkey, howler: *Alouatta palliata pigra*—Belize

Monkey, langur: *Presbytis entellus*

Monkey, silvered leaf: *Trachypithecus cristatus* (see lutung; langur)

Monkey, spider: *Ateles geoffroyi*—Tikal, Guatemala

Monkey, squirrel: *Saimiri* spp

Monkey, ursine colobus: *Colobus vellerosus*

Muriqui: *Brachyteles arachnoides*

Orangutan: *Pongo pygmaeus*

Pochard, common: *Aythya ferina*

Prairie dog, black-tailed: *Cynomys ludovicianus*

Sea lion, Steller: *Eumetopias jubatus*

Seal, elephant: *Miroungaangustirostris*

Seal, hooded: *Cystophora cristata*

Siamang: *Symphalangus syndactylus* (formerly *Hylobates syndactylus*)

Sifaka, Milne-Edwards: *Propithecus edwardsi*

Sparrow, house: *Passer domesticus*

Tamarin, saddle-back: *Saguinus fuscollis*

Walrus: *Odobenus rosmarus*

Bibliography

Adams, E., & Burnett, G. (1991). Scientific vocabulary divergence among female primatologists working in East Africa. *Social Studies of Science*, *21*, 547–560.

Ahmad, A., & Mohamad, K. (1996). The socioemotional development of orphans in orphanages and traditional foster care in Iraqi Kurdistan. *Child Abuse and Neglect*, *20*, 1161–1174.

Aiello, L., & Key, C. (2002). Energetic consequences of being a *Homo erectus* female. *American Journal of Human Biology*, *14*, 551–565.

Ali, Y. (2008). Honor, the state, and its implications: An examination of honor killing in Jordan and the efforts of local activists. Auburn, AL: Auburn University.

Alio, A. P., Mbah, A. K., Kornosky, J. L., Wathington, D., Marty, P. J., & Salihu, H. M. (2011). Assessing the impact of paternal involvement on racial/ethnic disparities in infant mortality rates. *Journal of Community Health*, *36*, 63–68.

Amarasinghe, A. A. T., Botejue, W. M. S., & Harding, L. E. (2009). Social behaviours of captive *Trachypithecus cristatus* (Mammalia: Cercopithecidae) in the National Zoological Gardens of Sri Lanka. *Taprobanica*, *1*, 66–73.

Ambrose, S. (2001). Paleolithic technology and human evolution. *Science*, *291*, 1748.

Anema, A., Joffres, M. R., Mills, E., & Spiegel, P. B. (2008). Widespread rape does not directly appear to increase the overall HIV prevalence in conflict-affected countries: So now what? *Emerging Themes in Epidemiology*, *5*, 1742–1776.

Anestis, S. (2010). Hormones and social behavior in primates. *Evolutionary Anthropology*, *19*, 66–78.

Anzai, T., Shiina, T., Kimura, N., Yanagiya, K., Kohara, S., Shigenari, A., Yamagata, T., ... Inoko, H. (2003). Comparative sequencing of human and chimpanzee MHC class I regions unveils insertions/deletions as the major path to genomic divergence. *Proceedings of the National Academy of Sciences, 100*, 7708–7713.

Ardrey, R. (1961). *African genesis: A personal investigation into the animal origins and nature of man.* London: Collins.

Ardrey, R. (1966). *The territorial imperative: A personal inquiry into the animal origins of property and nations.* New York: Atheneum.

Ardrey, R. (1976). *The hunting hypothesis: A personal conclusion concerning the evolutionary nature of man.* New York: Atheneum.

Argue, D., Donlon, D., Groves, C., & Wright, R. (2006). *Homo floresiensis*: Microcephalic, pygmoid, *Australopithecus*, or *Homo*? *Journal of Human Evolution, 51*, 360–374.

Aureli, F., & Schaffner, C. (2008). Social interactions, social relationships and the social system of spider monkeys. In C. Campbell (Ed.), *Spider monkeys: Behavior, ecology and evolution of the genus* Ateles (pp. 236–265). Cambridge: Cambridge University Press.

Bagemihl, B. (1999). *Biological exuberance: Animal homosexuality and natural diversity.* New York: St. Martin's Press.

Bales, K. (1980). Cumulative scaling of paternalistic behavior in non-human primates. *American Naturalist, 116*, 454–461.

Barrett, L., Gaynor, D., & Henzi, S. P. (2002). A dynamic interaction between aggression and grooming reciprocity among female chacma baboons. *Animal Behaviour, 63*(6), 1047–1053.

Bartlett, T. (2003). Intragroup and intergroup social interactions in white-handed gibbons. *International Journal of Primatology, 24*, 239–259.

Bartlett, T. (2011). The Hylobatidae: Small apes of Asia. In C. Campbell, A. Fuentes, et al. (Eds.), *Primates in perspective* (2nd ed., pp. 300–312). New York: Oxford University Press.

Bartlett, T. Q., Sussman, R. W., & Cheverud, J. M. (1993). Infant killing in primates: A review of observed cases with specific reference to the sexual selection hypothesis. *American Anthropologist, 95*(4), 958–990.

Bekoff, M. (2006). *Animal passions and beastly virtues: Reflections on redecorating nature.* Philadelphia: Temple University Press.

Bekoff, M., & Pierce, J. (2009). *Wild justice: The moral lives of animals.* Chicago: University of Chicago Press.

Berman, C. M., Ionica, C., & Li, J. (2007). Supportive and tolerant relationships among male Tibetan macaques at Huangshan, China. *Behaviour, 144*, 631–661.

Bernstein, I. (1968). The lutong of Kuala Selangor. *Behaviour, 32*, 1–16.

Birkhead, T. (2000). *Promiscuity: An evolutionary history of sperm competition and sexual conflict.* London: Faber and Faber.

Björkqvist, K. (1994). Sex differences in physical, verbal, and indirect aggression: A review of recent research. *Sex Roles, 30*, 177–188.

Blumstein, D. (2000). The evolution of infanticide in rodents: a comparative analysis. In C. van Schaik & C. Janson (Eds.), *Infanticide by males and its implications* (pp. 178–197). Cambridge: Cambridge University Press.

Boesch, C. (2002). Behavioural diversity in *Pan.* In C. Boesch, G. Hohmann, et al. (Eds.), *Behavioural diversity in chimpanzees and bonobos* (pp. 1–8). Cambridge: Cambridge University Press.

Boesch, C. (2009). *The real chimpanzee: Sex strategies in the forest.* Cambridge: Cambridge University Press.

Bolin, I. (1981). Male parental behavior in black howler monkeys (*Alouatta palliata pigra*) in Belize and Guatemala. *Primates, 22*(3), 349–360.

Borries, C., Launhardt, K., Epplen, C., Epplen, J. T., & Winkler, P. (1999). Males as infant protectors in hanuman langurs (*Presbytis entellus*) living in multimale groups—defence pattern, paternity and sexual behaviour. *Behavioral Ecology and Sociobiology, 46*, 350–356.

Bowles, S., & Gintis, H. (2011). *A co-operative species: Human reciprocity and its evolution.* Princeton, NJ: Princeton University Press.

Bradshaw, G. (2010). You see me, but do you hear me? The science and sensibility of trans-species dialogue. *Feminism and Psychology, 20*, 407–419.

Bradshaw, G., & Schore, A. (2007). How elephants are opening doors: Developmental neuroethology, attachment and social context. *Ethology, 113*, 426–436.

Bradshaw, G. A., Schore, A. N., Brown, J. L., Poole, J. H., & Moss, C. J. (2005). Concept elephant breakdown. *Nature, 433*, 807.

Bramble, D., & Lieberman, D. (2004). Endurance running and the evolution of *Homo*. *Nature, 432*, 345–352.

Brener, N. D., McMahon, P. M., Warren, C. W., & Douglas, K. A. (1999). Forced sexual intercourse and associated health-risk behaviors among female college students in the United States. *Journal of Consulting and Clinical Psychology, 67*, 252.

Brockelman, W. Y. (2009). Ecology and the social system of gibbons. In D. Whittaker & S. Lappan (Eds.), *The gibbons: New perspectives on small ape socioecology and population biology* (pp. 211–239). New York: Springer.

Bronikowski, A. M., Altmann, J., Alberts, S. C., Brockman, D. K., Cords, M., Fedigan, L. M., Pusey, A., ... Strier, K. B. (2011). Aging in the natural world: Comparative data reveal similar mortality patterns across primates. *Science, 331*, 1325–1328.

Brunner, H., Nelen, M., Breakefield, X. O., Ropers, H. H., & van Oost, B. A. (1993). Abnormal behavior associated with a point mutation in the structural gene for monoamine oxidase A. *Science, 262*, 578–580.

Byrne, W., & Whiten, A. (1988). *Machiavellian intelligence*. Oxford: Oxford University Press.

Campbell, B. C., Dreber, A., Apicella, C. L., Eisenberg, D. T. A., Gray, P. B., Little, A. C., Garcia, J. R., ... Lum, J. K. (2010). Testosterone exposure, dopaminergic reward, and sensation-seeking in young men. *Physiology and Behavior, 99*, 451–456.

Catani, C., Schauer, E., Elbert, T., Missmahl, I., Bette, J. P., & Neuner, F. (2009). War trauma, child labor, and family violence: Life adversities and PTSD in a sample of school children in Kabul. *Journal of Traumatic Stress*, *22*, 163–171.

Chance, M. (1956). Social behaviour of rhesus monkey: Social structure of a colony of *Macaca mulatta*. *British Journal of Animal Behaviour*, *4*, 1–13.

Chang, I. (1997). *The rape of Nanking: The forgotten holocaust of World War II.* New York: BasicBooks.

Chapais, B., & Berman, C. (Eds.). (2004). *Kinship and behavior in primates*. New York: Oxford University Press.

Cheney, D., & Seyfarth, R. (2007). *Baboon metaphysics: The evolution of a social mind.* Chicago: University of Chicago Press.

Cherkas, L. F., Oelsner, E. C., Mak, Y., Valdes, A., & Spector, T. D. (2004). Genetic influences on female infidelity and number of sexual partners in humans: A linkage and association study of the role of the vasopressin receptor gene (AVPR1A). *Twin Research*, *7*, 649–658.

Chivers, D. (1974). *The Siamang in Malaya: A field study of a primate in tropical rain forest*. Basel: S. Karger.

Chivers, D. (1984). Gibbons. In D. MacDonald (Ed.), *The encyclopedia of mammals* (pp. 376–380). New York: Facts on File.

Chivers, D. (2001). The swinging singing apes: Fighting for food and family in far-east forests. In *The apes: Challenges for the 21st century* (pp. 1–28). Chicago: Chicago Zoological Society, Brookfield Zoo, May 10–13, 2000.

Ciochon, R. L., & Corruccini, R. S. (1977). The coraco-acromial ligament and projection index in man and other anthropoid primates. *Journal of Anatomy, 124*(Pt 3), 627–632.

Clutton-Brock, T. (2007). *Meerkat Manor: Flower of the Kalahari.* London: Phoenix.

Cluver, L., & Gardner, F. (2006). The psychological well-being of children orphaned by AIDS in Cape Town, South Africa. *Annals of General Psychiatry*, *5*, 8.

Cluver, L., & Gardner, F. (2007). The mental health of children orphaned by AIDS: A review of international and southern African research. *Journal of Child and Adolescent Mental Health, 19*, 1–17.

Cohen, J. (2010). *Almost chimpanzee: Searching for what makes us human, in rainforests, labs, sanctuaries, and zoos*. New York: Times Books, Henry Holt.

Cowan, J. (2006). Are tarsiers really a taxonomic enigma? *Lambda Alpha Journal* (Washington State University), *36*, 18–29.

Coupé, C. & Hombert, J.-M. (2005). Polygenesis of linguistic strategies: A scenario for the emergence of languages. In J. W. Minett & W. S.-Y. Wang (Eds.), *Language Acquisition, Change and Emergence: essays in evolutionary linguistics* (pp. 153). Hong Kong: City University of Hong Kong Press.

Cuvier, F. G. (1819). *Histoire naturelle des mammifères*, Vol. 1, Livr. 6:1. Paris: A. Belin.

Dagg, A. (1978). *Camel quest: Summer research on the Saharan Camel.* Waterloo, ON: Otter Press.

Dagg, A. (1983). *Harems and other horrors: Sexual bias in behavioral biology.* Waterloo, ON: Otter Press.

Dagg, A. (1984). Homosexual behavior and female-male mounting in mammals—a first survey. *Mammal Review, 14*, 155–185.

Dagg, A. (1998). Infanticide by male lions hypothesis: A fallacy influencing research into human behavior. *American Anthropologist, 100*(4), 1–11.

Dagg, A. (2000). The infanticide hypothesis: A response to the response. *American Anthropologist, 102*(4), 831–834.

Dagg, A. (2005). *"Love of shopping" is not a gene: Problems with Darwinian psychology*. Montreal: Black Rose Press.

Dagg, A. (2009). *The social behavior of older animals*. Baltimore, MD: Johns Hopkins University Press.

Dagg, A. (2011). *Animal friendships*. Cambridge: Cambridge University Press.

Dallaire, R. (2003). *Shake hands with the devil: The failure of humanity in Rwanda.* Toronto: Random House.

Darwin, C. (1859). *On the origin of species*. London: John Murray.

Darwin, C. (1871). *The descent of man and selection in relation to sex*. London: John Murray.

Dawkins, R. (2006). *The selfish gene*. Oxford: Oxford University Press.

Deag, J., & Crook, J. (1971). Social behaviour and "agonistic buffering" in the wild barbary macaque *Macaca sylvana* L. *Folia Primatologica, 15*, 183–200.

Delacour, J. (1933). On the Indochinese gibbon (*Hylobates concolor*). *Journal of Mammalogy, 14*, 71–72.

Delgado, R., Jr., & Van Schaik, C. (2000). The behavioral ecology and conservation of the orangutan (*Pongo pygmaeus*): A tale of two islands. *Evolutionary Anthropology: Issues, News, and Reviews, 9*, 201–218.

de Waal, F. (1989). *Peacemaking among primates.* Cambridge, MA: Harvard University Press.

de Waal, F. (1997). *Bonobo: The forgotten ape.* Berkeley: University of California Press.

de Waal, F. (1998). Comment on article by Craig B. Stanford. *Current Anthropology, 39*(4), 407–408.

Diamond, J. (1992). *The third chimpanzee: The evolution and future of the human animal.* New York: Harper.

Diamond, J. (1997). *Why sex is fun: The evolution of human sexuality.* London: Weidenfeld and Nicolson.

Dielentheis, T. F., Zaiss, E., & Geissmann, T. (1991). Infant care in a family of siamangs (*Hylobates syndactylus*) with twin offspring at Berlin Zoo. *Zoo Biology, 10*, 309–317.

Draper, P. (1978). The learning environment for aggression and anti-social behavior among the !Kung. In Ashley Montagu (Ed.), *Learning non-aggression: The experience of non-literate societies* (pp. 31–93). New York: Oxford University Press.

Dreber, A., Rand, D. G., Garcia, J. R., Wernerfelt, N., Lum, J. K., & Zeckhauser, R. (2010). *Dopamine and risk preferences in different domains* (Harvard Kennedy School Faculty Research Working Paper Series RWP10-012, 21).

Dubois, M. (2010, Summer). Aid dilemma in Papua New Guinea. *Dispatches: Médecins sans Frontières Magazine*, 10–11.

Dunbar, R. (1991). Functional significance of social grooming in primates. *Folia Primatologica, 57*, 121–131.

Dunbar, R. (1995). The mating system of callitrichid primates: I. Conditions for the coevolution of pair bonding and twinning. *Animal Behaviour, 50*, 1057–1070.

Dunbar, R. (1996). *Grooming, gossip, and the evolution of language.* Cambridge, MA: Harvard University Press.

Dunbar, R. (2003). The social brain: Mind, language, and society in evolutionary perspective. *Annual Review of Anthropology, 32*, 163–181.

Edgar, B. (2010). The power of chocolate. *Archaeology*, 63. Retrieved from http://www.archaeology.org/1011/abstracts/chocolate.html

Eisenberg, D., A., Campbell, B. C., MacKillop, J., Modi, M., Dang, D., Lum, J. K., & Wilson, D. S. (2007). Polymorphisms in the dopamine D4 and D2: Receptor genes and reproductive and sexual behaviors. *Evolutionary Psychology*, *5*, 696–715.

Eisenberg, D. T. A., Apicella, C. L., Campbell, B. C., Dreber, A., Garcia, J. R., & Lum, J. K. (2009). Assortative human pair-bonding for partner ancestry and allelic variation of the dopamine receptor D4 (DRD4) gene. *Social Cognitive and Affective Neuroscience, 5*(2–3), 194–202.

Elder, A. (2009). Hylobatid diets revisited: The importance of body mass, fruit availability, and interspecific competition. In D. W. Lappan & S. Lappan (Eds.), *The gibbons: New perspectives on small ape socioecology and population biology* (pp. 133–159). New York: Springer.

Ellis, S. (2003). *The journey of man: A genetic odyssey*. New York: Random House.

Eriksson, A. & Manica, A. (2012). Effect of ancient population structure on the degree of polymorphism shared between modern human populations and ancient hominins. *Proceedings of the National Academy of Sciences*, (published on-line before print, 14 August 2012.

Estrada, A. (1984). Male-infant interactions among free-ranging stumptail macaques. In D. Taub (Ed.), *Primate paternalism* (pp. 56–87). New York: Van Nostrand Reinhold.

Fairbanks, L. (1990). Reciprocal benefits of allomothering for female vervet monkeys. *Animal Behaviour*, *40*(3), 553–562.

Fan, P.-f., Jiang, X. L., Liu, C. M., & Luo, W. S. (2006). Polygynous mating system and behavioural reason of black crested gibbon (*Nomascus concolor jingdongensis*) at Dazhaizi, Mt. Wuliang, Yunnan, China. *Zoological Research*, *27*, 216–220.

Fedigan, L. (1982). *Primate paradigms: Sex roles and social bonds.* Montreal: Eden Press.

Fedigan, L., & Baxter, M. (1984). Sex differences and social organization in free-ranging spider monkeys (*Ateles geoffroyi*). *Primates*, *25*(3), 279–294.

Fernandez-Duque, E. (2007). Aotinae: Social monogamy in the only nocturnal haplorhines. In C. Campbell, S. Bearder, et al. (Eds.), *Primates in perspective* (pp. 139–154). Oxford: Oxford University Press.

Fernandez-Duque, E., Valeggia, C. R., & Mendoza, S. P. (2009). The biology of paternal care in human and nonhuman primates. *Annual Review of Anthropology*, *38*, 115–130.

Fietz, J. (1999). Monogamy as a rule rather than exception in nocturnal lemurs: The case of the fat-tailed dwarf lemur, *Cheirogaleus medius. Ethology*, *105*, 255–272.

Firth, R. (1957). *We, the Tikopia: A sociological study of kinship in primitive Polynesia.* Palo Alto, CA: Stanford University Press.

Flannery, T. (2010). *Here on Earth: A natural history of the planet.* Toronto: HarperCollins.

Flinn, M. V. (2011). Evolutionary anthropology of the human family. In C. Salmon & T. K. Shackelford (Eds.), *The Oxford Handbook of Evolu-*

tionary Family Psychology (pp. 12–32). New York: Oxford University Press.

Fossey, D. (1983). *Gorillas in the mist.* Boston: Houghton Mifflin.

Foster, K., Wenseleers, T., & Ratnieks, F. L.W. (2006). Kin selection is the key to altruism. *Trends in Ecology and Evolution, 21*, 57–60.

Fry, D. (2006). *The human potential for peace: An anthropological challenge to assumptions about war and violence.* New York: Oxford University Press.

Fuentes, A. (1998). Re-evaluating primate monogamy. *American Anthropologist*, New Series, *100*(4), 890–907.

Furuichi, T., & Ihobe, H. (1994). Variation in male relationships in bonobos and chimpanzees. *Behaviour, 130*(3–4), 211–228.

Galdikas, B. (1985). Subadult male orangutan sociality and reproductive behavior at Tanjung Putting. *American Journal of Primatology, 8*, 87–99.

Galdikas, B. (1988). Orangutan diet, range, and activity at Tanjung Puting, Central Borneo. *International Journal of Primatology, 9*, 1–35.

Gee, H. (2004). *Jacob's Latter: the history of the human genome.* New York, London: W.W. Norton & Company.

Geissmann, T., & Anzenberger, G. (2009). Hormonal correlates of the ovarian cycle in the yellow-cheeked crested gibbon (*Nomascus gabriellae*), and a review of ovarian cycles in gibbons (Hylobatidae). *Gibbon Journal, 5*, 61–73.

Geist, V. (1975). *Mountain sheep and man in the northern wilds.* Ithaca, NY: Cornell University Press.

Geoffroy Saint-Hilaire, É., and Cuvier, F.G. 1819. *Histoire Naturelle des Mammifères*. A Belin, Paris.

Gettler, L. (2010). Direct male care and hominin evolution: Why male-child interaction is more than a nice social idea. *American Anthropologist, 112*, 7–21.

Gettler, L. T., Mcdade, T. W., & Kuzawa, C. W. (2011a). Cortisol and testosterone in Filipino young adult men: Evidence for co-regulation of both hormones by fatherhood and relationship status. *American Journal of Human Biology, 23*, 609–620.

Gettler, L. T., McDade, T. W., Feranil, A. B., & Kuzawa, C. W. (2011b). Longitudinal evidence that fatherhood decreases testosterone in human males. *Proceedings of the National Academy of Sciences, 108*(39), 16194–16199.

Gibbons, A. (2011, August 26). Who were the Denisovans? *Science, 333*(6046), 1084–1087.

Gittins, S., & Raemaekers, J. (1980). Siamang, lar and agile gibbons. In D. Chivers (Ed.), *Malayan forest primates: Ten years' study in tropical rain forest* (pp. 63–105). New York: Plenum Press.

Goldizen, A. (1990). A comparative perspective on the evolution of tamarin and marmoset social systems. *International Journal of Primatology, 11*(1), 63–83.

Goldizen, A. (2003). Social monogamy and its variations in callitrichids: Do these relate to the costs of infant care? In U. Reichard & C. Boesch (Eds.), *Monogamy: Mating strategies and partnerships in birds, humans and other mammals* (pp. 232–247). Cambridge: Cambridge University Press.

Goodall, J. (1965). Chimpanzees of the Gombe Stream Reserve. In I. DeVore (Ed.), *Primate behavior: Field studies of monkeys and apes* (pp. 425–473). New York: Holt, Rinehart and Winston.

Goodall, J. (1986). *The chimpanzees of Gombe: Patterns of behavior.* Cambridge, MA: Belknap Press.

Goodman, M., Grossman, L., and Wildman, D.E. (2005). Moving primate genomics beyond the chimpanzee genome. *Trends in Genetics, 21*, 511–517.

Goodman, M., Porter, C. A., Czelusniak, J., Page, S. L., Schneider, H., Shoshani, J., Gunnell, G., ... Groves, C. P. (1998). Toward a phylogenetic classification of primates based on DNA evidence complemented by fossil evidence. *Molecular Phylogenetics and Evolution, 9*, 585–598.

Goodman, M., Tagle, D. A., Fitch, D. H., Bailey, W., Czelusniak, J., Koop, B. F., Benson, P., ... Slightom, J. L. (1990). Primate evolution at the

DNA level and a classification of hominoids. *Journal of Molecular Evolution, 30*, 260–266.

Gribbin, J., & Cherfas, J. (2001). *The first chimpanzee: In search of human origins.* London: Penguin Books.

Hanson, T. (2008). *The impenetrable forest: My gorilla years in Uganda.* Warwick, NY: 1500 Books.

Harcourt, A. (1979). Social relationships between adult male and female mountain gorillas in the wild. *Animal Behaviour, 27*, 325–342.

Harcourt, A. H., Harvey, P. H., Larson, S., & Short, R. (1981). Testis weight, body weight and breeding system in primates. *Nature, 293*(3), 55–57.

Harding, L. (2010). *Trachypithecus cristatus. Mammalian Species, 42*(1), 149–165.

Harding, L. E. (2011). Unusual Affiliative Behaviour in Orangutans. *Taprobanica, 3*(2), 110–111.

Harding, L. E. (2011a). *Trachypithecus delacouri* (Primates: Cercopithecidae). *Mammalian Species, 43*(880), 118–128.

Harding, L. E. (2011b). Red morph of silvered lutung (*Trachypithecus cristatus*) rediscovered in Borneo, Malaysia. *Taprobanica, 3*(1), 47–48

Harding, L. E. (2011c). Rare mammals recorded in Borneo, Malaysia. *Taprobanica, 3*(2), 107–109.

Harding, L. E. (2012). *Nomascus leucogenys. Mammalian Species, 44*(890), 1–15.

Hart, D., & Sussman, R. (2005). *Man the hunted: Primates, predators, and human evolution.* New York: Westview Press.

Hart, D., & Sussman, R. (2009). *Man the hunted: Primates, predators, and human evolution* (expanded ed.). Philadelphia: Westview Press.

Hasegawa, T., & Hiraiwa, M. (1980). Social interactions of orphans observed in a free-ranging troop of Japanese monkeys. *Folia Primatologica, 33*, 129–158.

Hawkes, K. (2003). Grandmothers and the evolution of human longevity. *American Journal of Human Biology, 15*, 380–400.

Henkel, S., Heistermann, M., & Fischer, J. (2010). Infants as costly social tools in male Barbary macaque networks. *Animal Behaviour*, *79*, 1199–1204.

Hesketh, T. (2009). Too many males in China: The causes and the consequences. *Significance*, *6*, 9–13.

Hewlett, B., & MacFarlan, S. (2010). Fathers' role in hunter-gatherers and other small-scale cultures. In M. Lamb (Ed.), *The role of the father in child development* (5th ed., pp. 413–434). Hoboken, NJ: John Wiley.

Hill, D. (1994). Affiliative behaviour between adult males of the genus *Macaca*. *Behaviour*, *130*(3–4), 293–308.

Hill, D., & van Hooff, J. (1994). Affiliative relationships between males in groups of nonhuman primates: A summary. *Behaviour*, *130*(3–4), 143–149.

Hoogland, J. (1994). Nepotism and infanticide among prairie dogs. In S. Parmigiani & F. vom Saal (Eds.), *Infanticide and parental care* (pp. 321–337). Chur, Switzerland: Harwood Academic Publishers.

Horgan, J. (2010, June 29). Quitting the hominid fight club: The evidence is flimsy for innate chimpanzee—let alone human—warfare. Retrieved from http://www.scientificamerican.com blog

Horner, V., Carter, J. D., Suchak, M., & de Waal, F. (2011). Spontaneous prosocial choice by chimpanzees. *Proceedings of the National Academy of Sciences*, *108*(33), 13847–13851.

Hrdy, S. (1974). Male-male competition and infanticide among the langurs (*Presbytis entellus*) of Abu, Rajasthan. *Folia Primatologica*, *22*, 19–58.

Hrdy, S. (1976). Care and exploitation of nonhuman primate infants by conspecifics other than the mother. In J. Rosenblatt (Ed.), *Advances in the study of behavior* (pp. 101–158). New York: Academic Press.

Hrdy, S. (1979). Infanticide among animals: A review, classification, and examination of the implications for the reproductive strategies of females. *Ethology and Sociobiology*, *1*, 13–40.

Hrdy, S. (1992). Fitness tradeoffs in the history and evolution of delegated mothering with special reference to wet-nursing, abandonment, and infanticide. *Ethology and Sociobiology*, *13*, 409–442.

Hrdy, S., & Hrdy, D. (1976). Hierarchical relations among female hanuman langurs (Primates: Colobinae, *Presbytis entellus*). *Science*, *193*, 913.

Huxley, T. H. (1870). *Biogenesis and abiogenesis*. London: Presidential Address at the British Association (later published in *Collected Essays*, Vol. 8, p. 229).

Itani, J. (1959). Paternal care in the wild Japanese monkey, *Macaca fuscata fuscata*. *Primates*, *2*, 61–93.

Jablonski, N. G., & Chaplin, G. (1993). Origin of habitual terrestrial bipedalism in the ancestor of the Hominidae. *Journal of Human Evolution, 24*(4), 259–280.

Jahme, C. (2001). *Beauty and the beasts: Woman, ape and evolution.* New York: Soho Press.

Joffe, T. (1997). Social pressures have selected for an extended juvenile period in primates. *Journal of Human Evolution*, *32*, 593–605.

Johnson, A., & Earle, T. (2000). *The evolution of human societies* (2nd ed.). Palo Alto, CA: Stanford University Press.

Johnson, S., Bodon, J., & McCroskey, J. C. (2008). Nature nor nurture: Understanding verbal aggression in the military. *Human Communication, 11*, 115–132.

Kano, T. (1992). *The last ape: Pygmy chimpanzee behavior and ecology*. Palo Alto, CA: Stanford University Press.

Kaplan, H., Hill, K., Lancaster, J., & Hurtado, A. M. (2000). A theory of human life history evolution: Diet, intelligence, and longevity. *Evolutionary Anthropology: Issues, News, and Reviews*, *9*, 156–185.

Kappeler, P., & Van Schaik, C. (2002). Evolution of primate social systems. *International Journal of Primatology*, *23*, 707–740.

Kosfeld, M., Heinrichs, M., Zak, P. J., Fischbacher, U., & Fehr, E. (2005). Oxytocin increases trust in humans. *Nature*, *433*, 673–676.

Koyama, N. (1973). Dominance, grooming, and clasped-sleeping relationships among bonnet monkeys in India. *Primates, 14*(2–3), 225–244.

Kudo, H., & Dunbar, R. (2001). Neocortex size and social network size in primates. *Animal Behaviour, 62*, 711–722.

Kummer, H. (1967). Tripartite relations in hamadryas baboons. In S. Altmann (Ed.), *Social communication among primates* (pp. 63–71). Chicago: University of Chicago Press.

Kummer, H. (1995). *In quest of the sacred baboon: A scientist's journey.* Princeton, NJ: Princeton University Press.

Kuzawa, C. W., Gettler, L. T., Muller, M. N., McDade, T. W., & Feranil, A. B. (2009). Fatherhood, pairbonding and testosterone in the Philippines. *Hormones and Behavior, 56*, 429–435.

Lappan, S. (2008). Male care of infants in a siamang (*Symphalangus syndactylus*) population including socially monogamous and polyandrous groups. *Behavioral Ecology and Sociobiology, 62*, 1307–1317.

Leakey, L., & Ardrey, R. (1971). Dialogue on aggression and violence in man. *Munger Africana Library*, II (9):1–25.

Le Boeuf, B., & Campagna, C. (1994). Protection and abuse of young in pinnipeds. In S. Parmigiani and F. vom Saal (Eds.), *Infanticide and Parental Care* (pp. 257–276). Switzerland: Harwood Academic Publishers.

Lee, R., & DeVore, I. (Eds.). (1968). *Man the hunter*. Chicago: Aldine Publishing.

Leffler, A., Gillespie, D. L. & Conaty, J. C. (1982). The effects of status differentiation on nonverbal behavior. *Social Psychology Quarterly, 45*(3), 153–161.

Lesson, R.-P. (1827). *Manuel de mammalogie ou histoire naturelle des mammifères*. Paris: Libraire de Roret.

Leutenegger, W., & Kelly, J. (1977). Relationship of sexual dimorphism in canine size and body size to social, behavioral, and ecological correlates in anthropoid primates. *Primates, 18*, 117–136.

Linnaeus, C. (1758). *Systema naturae.* Holmiae [1–4] Tomus I., Laurentii Salvii.

Lorenz, K. (1966). *On aggression.* London: Methuen.

Lovejoy, C. O., Simpson, S. W., White, T. D., Asfaw, B., & Suwa, G. (2009). Careful climbing in the Miocene: The forelimbs of *Ardipithecus ramidus* and humans are primitive. *Science*, *326*, 70.

Lovejoy, C. O., Suwa, G., Simpson, S. W., Matternes, J. H., & White, T. D. (2009). The great divides: *Ardipithecus ramidus* reveals the postcrania of our last common ancestors with African apes. *Science, 326*, 100–106.

Macedo, R., & Bianchi, C. (1997). When birds go bad: Circumstantial evidence for infanticide in the communal South-American guira cuckoo. *Ethology, Ecology and Evolution*, *9*, 45–54.

Maggioncalda, A. N., & Sapolsky, R. M. (2002). Disturbing behaviors of the orangutan. *Scientific American, 286*(6), 60–65.

Maples Jr, E., Haraway, M., & Hutto, C. (1989). Development of coordinated singing in a newly formed siamang pair (*Hylobates syndactylus*). *Zoo Biology*, *8*(4), 367–378.

Marlowe, F. W. (2003). The mating system of foragers in the Standard Cross-Cultural Sample. *Cross-Cultural Research, 37*(3), 282–306.

Martin, W. (1841). *A general introduction to the natural history of mammiferous animals: With a particular view of the physical history of man and the more closely allied genera of quadrumanes, or monkeys.* London: Wright and Co.

Mayell, H. (2002, February 12). Thousands of women killed for family "honor." *National Geographic News*, 1–3.

McCarthy, S. (2004). *Becoming tiger: How baby animals learn to live in the wild.* New York: HarperCollins.

McGrew, W. (2010). New theaters of conflict in the animal culture wars: Recent findings from chimpanzees. In E. Lonsdorf, S. Ross, et al. (Eds.), *The mind of the chimpanzee: Ecological and experimental perspectives* (pp. 168–177). Chicago: University of Chicago Press.

McPherron, S. P., Alemseged, Z., Marean, C. W., Wynn, J. G., Reed, D., Geraads, D., Bobe, R., ... Bearat, H. A. (2010). Evidence for stone-tool-assisted consumption of animal tissues before 3.39 million years ago at Dikika, Ethiopia. *Nature*, *466*, 857–860.

Meggitt, M. (1977). *Blood is their argument: Warfare among the Mae Enga tribesmen of the New Guinea Highlands.* Palo Alto, CA: Mayfield Publishing.

Michod, R. (1982). The theory of kin selection. *Annual Review of Ecology and Systematics*, *13*, 23–55.

Montagu, A. (1978). Introduction. In Ashley Montagu (Ed.), *Learning non-aggression: The experience of non-literate societies* (pp. 3–11). New York: Oxford University Press.

Morelli, T. L., King, S. J., Pochron, S. T., & Wright, P. C. (2009). The rules of disengagement: Takeovers, infanticide, and dispersal in a rainforest lemur, *Propithecus edwardsi. Behaviour*, *146*, 499–523.

Morris, D. (1967). *The naked ape: A zoologist's study of the human animal.* New York: McGraw-Hill.

Morwood, M., Soejono, R., et al. (2004). Archaeology and age of a new hominin from Flores in eastern Indonesia. *Nature*, *431*, 1087–1091.

Mukwege, M., & Nangini, C. (2009). Rape with extreme violence: The new pathology in South Kivu, Democratic Republic of Congo. *PLoS Medicine*, *6*, e1000204.

Nishida, T. (1990). *The chimpanzees of the Mahale Mountains: Sexual and life history strategies.* Tokyo: University of Tokyo Press.

Norton, G. W., Rhine, R. J., Wynn, G. W., & Wynn, R. D. (1987). Baboon diet: A five-year study of stability and variability in the plant feeding and habitat of the yellow baboons (*Papio cynocephalus*) of Mikumi National Park, Tanzania. *Folia Primatologica*, 48, 78–120.

O'Connell, J. F., Hawkes, K., & Blurton Jones, N. G. (1999). Grandmothering and the evolution of *Homo erectus*. *Journal of Human Evolution*, *36*, 461–485.

O'Connor, D. B., Archer, J., Hair, W. M., & Wu, F. C. W. (2002). Exogenous testosterone, aggression, and mood in eugonadal and hypogonadal men. *Physiology and Behavior*, *75*, 557–566.

Ogawa, H. (2006). *Wily monkeys: Social intelligence of Tibetan macaques*. Kyoto: Kyoto University Press.

Packer, C. (1979). Inter-troop transfer and inbreeding avoidance in *Papio anubis*. *Animal Behaviour*, *27*, 1–36.

Packer, C. (1994). *Into Africa*. Chicago: University of Chicago Press.

Packer, C. (2000). Infanticide is no fallacy. *American Anthropologist*, *102*, 829–831.

Pagel, M., & Bodmer, W. (2003). A naked ape would have fewer parasites. *Proceedings of the Royal Society B (Suppl.)*, *270*, S117–S119.

Palombit, R. (1995). Longitudinal patterns of reproduction in wild female siamang (*Hylobates syndactylus*) and white-handed gibbons (*Hylobates lar*). *International Journal of Primatology*, *16*, 739–760.

Palombit, R. (2009). "Friendship" with males: A female counterstrategy to infanticide in chacma baboons of the Okavango Delta. In M. Muller & R. Wrangham (Eds.), *Sexual coercion in primates and humans: An evolutionary perspective on male aggression against females* (pp. 377–409). Cambridge, MA: Harvard University Press.

Panksepp, J. (1998). *Affective neuroscience: The foundations of human and animal emotions*. New York: Oxford University Press.

Paoli, T. (2009). The absence of sexual coercion in bonobos. In M. Muller and R. Wrangham (Eds.), *Sexual coercion in primates and humans: An evolutionary perspective on male aggression against females* (pp. 410–423). Cambridge, MA: Harvard University Press.

Paul, A., Preuschoft, S., & van Schaik, C. P. (2000). The other side of the coin: Infanticide and the evolution of affiliative male-infant interactions in Old World primates. In C. van Schaik & C. Janson (Eds.), *Infanticide by males and its implications* (pp. 269–292). Cambridge: Cambridge University Press.

Peterson, D. (2006). *Jane Goodall: The woman who redefined man*. Boston: Houghton Mifflin.

Pickering, R., Woodhead, J. D., Hellstrom, J. C., Dirks, P. H. G. M., Jinnah, Z., Berger, L. R., De, R. D. J., ... Herries, A. I. R. (2011). *Australopithecus sediba* at 1.977 Ma and implications for the origins of the genus *Homo*. *Science, 333,* 6048, 1421–1423.

PLoS Medicine Editors. (2009). Rape in war is common, devastating, and too often ignored. *PLoS Medicine*, 6.e1000021

Poirier, F. (1974). Colobine aggression: A review. In R. L. Holloway (Ed.), *Primate aggression, territoriality, and xenophobia* (pp. 123–157). New York: Academic Press.

Pollard, K. (2009). What makes us human? Comparisons of the genomes of humans and chimpanzees are revealing those rare stretches of DNA that are ours alone. *Scientific American, 300*, 44–49.

Pook, A. (1984). The evolutionary role of socio-ecological factors in the development of paternal care in the New World family Callitrichidae. In D. Taub (Ed.), *Primate paternalism* (pp. 336–376). New York: Van Nostrand Reinhold.

Power, M. (1991). *The egalitarians—Human and chimpanzee: An anthropological view of social organization.* Cambridge: Cambridge University Press.

Prokop, P., Trnka, R., & Trnka, A. (2009). First videotaped infanticide in the common pochard *Aythya ferina. Biologia*, *64*(5), 1016–1017.

Provan, I. W., Long, V. P., & Longman, T. (2003). *A biblical history of Israel*. Louisville, KY: Westminster John Knox Press.

Raffaele, P. (2010). *Among the great apes: Adventures on the trail of our closest relatives.* New York: Smithsonian Books.

Rantala, M. (2007). Evolution of nakedness in *Homo sapiens*. *Journal of Zoology*, *273*, 1–7.

Reese, A. (2009). *The infanticide controversy: Primatology and the art of field science*. Chicago: University of Chicago Press.

Reichard, U. (1995). Extra-pair copulations in a monogamous gibbon (*Hylobates lar*). *Ethology*, *100*, 99–112.

Reichard, U. (2003a). Monogamy: Past and present. In U. Reichard & C. Boesch (Eds.), *Monogamy: Mating strategies and partnerships in birds, humans and other mammals* (pp. 3–25). Cambridge: Cambridge University Press.

Reichard, U. (2003b). Social monogamy in gibbons: The male perspective. In U. Reichard & C. Boesch (Eds.), *Monogamy: Mating strategies and partnerships in birds, humans and other mammals* (pp. 190–213). Cambridge: Cambridge University Press.

Reichard, U., & Sommer, V. (1997). Group encounters in wild gibbons (*Hylobates lar*): Agonism, affiliation, and the concept of infanticide. *Behaviour, 134*, 1135–1174.

Rennie, J. (1838). *The menageries: The natural history of monkeys, opossums, and lemurs* (Vol. 1). London: Nattali and Bond.

Reno, P. L., Meindl, R. S., McCollum, M. A., & Lovejoy, C. O. (2003). Sexual dimorphism in *Australopithecus afarensis* was similar to that of modern humans. *Proceedings of the National Academy of Sciences, 100*, 9404–9409.

Reynolds, V. (1975). How wild are the Gombe chimpanzees? *Man* [new series], *10*, 123–125.

Reynolds, V. (2005). *The chimpanzees of the Budongo Forest: Ecology, behaviour, and conservation.* Oxford, UK: Oxford University Press.

Richardson, S. (1993, October 1). A violence in the blood. *Discover Magazine.* Retrieved from http://discovermagazine.com/1993/oct/aviolenceinthebl293.

Richter, C., Mevis, L., Malaivijitnond, S., Schülke, O., & Ostner, J. (2009). Social relationships in free-ranging male *Macaca arctoides. International Journal of Primatology, 30*, 625–642.

Rightmire, G. (1993). *The evolution of* Homo erectus: *Comparative anatomical studies of an extinct human species.* Cambridge: Cambridge University Press.

Rilling, J. K. (2011). The neurobiology of cooperation and altruism. In R. W. Sussman & C. R. Cloninger (Eds.), *Origins of Cooperation and Altruism* (pp. 295–306). New York: Springer Verlag.

Robbins, M. (2009). Male aggression against females in mountain gorillas: Courtship or coercion? In M. Muller & R. Wrangham (Eds.),

Sexual coercion in primates and humans: An evolutionary perspective on male aggression against females (pp. 112–127). Cambridge, MA: Harvard University Press.

Roebroeks, W., & Villa, P. (2011). On the earliest evidence for habitual use of fire in Europe. *Proceedings of the National Academy of Sciences, 108*, 5209–5214.

Rogers, M. E., Abernethy, K., Bermejo, M., Cipolletta, C., Doran, D., McFarland, K., Nishihara, T., ... Tutin, C. E. (2004). Western gorilla diet: A synthesis from six sites. *American Journal of Primatology, 64*, 173–192.

Roonwal, M., & Mohnot, S. M. (1977). *Primates of South Asia: ecology, sociobiology, and behavior*. Cambridge, Massachusets and London, England: Harvard University Press.

Rossa, C., & Regan, G. (2000). Allocare, predation risk, social structure and natal coat colour in anthropoid primates. *Folia Primatologica, 71*, 67–76.

Roth, G., & Dicke, U. (2005). Evolution of the brain and intelligence. *Trends in Cognitive Sciences, 9*, 250–257.

Ryan, D. (2010, December 4, 2010). "Cheating gene" discovery sheds new light on human sexuality. *Vancouver Sun,* A9. Retrieved from http://www2.canada.com/vancouversun/news/westcoastnews/story.html?id = f03fa777-9ae0-407b-ad17-e4407f1031f7& p = 2.

Sade, D. (2000). The missing dyad. In Dean Falk (Ed.), *Primate diversity* (pp. 236–237). New York: Norton.

Sapolsky, R. (2001). *A primate's memoir.* New York: Scribner.

Sapolsky, R. (2011, Spring). Warrior baboons give peace a chance. *Yes!, 57*, 34–37.

Scelza, B. (2009). The grandmaternal niche: Critical caretaking among Martu Aborigines. *American Journal of Human Biology, 21*, 448–454.

Schaal, S., & Elbert, T. (2006). Ten years after the genocide: Trauma confrontation and posttraumatic stress in Rwandan adolescents. *Journal of Traumatic Stress, 19*, 95–105.

Schaller, G. (1972). *The Serengeti lion.* Chicago: University of Chicago Press.

Schmitt, D. (2003). Insights into the evolution of human bipedalism from experimental studies of humans and other primates. *Journal of Experimental Biology*, *206*, 1437.

Schülke, O. (2005). Evolution of pair-living in *Phaner furcifer. International Journal of Primatology*, *26*, 903–919.

Schultz, A. H. (1933). Observations on the growth, classification and evolutionar specializations of gibbons and siamangs. *Human Biology,* *5*, 212–255, 385–428.

Schulz, M. (2010, October 15). How Middle Eastern milk drinkers conquered Europe. *Der Schpiegel*. Retrieved from http://www.spiegel.de/international/zeitgeist/0,1518,723310-3,00.html

Schürmann, C., & Van Hooff, J. (1986). Reproductive strategies of the orangutan: New data and a reconsideration of existing sociosexual models. *International Journal of Primatology*, *7*, 265–287.

Scott, J. (2011). Folivory, frugivory, and postcanine size in the Cercopithecoidea revisited. *American Journal of Physical Anthropology*, *146*(1), 20–27.

Shea, J. (2006). The origins of lithic projectile point technology: Evidence from Africa, the Levant, and Europe. *Journal of Archaeological Science*, *33*, 823–846.

Shelmidine, N., Borries, C., & McCann, C. (2009). Patterns of reproduction in Malayan silvered leaf monkeys at the Bronx Zoo. *American Journal of Primatology*, *71*, 1–8.

Shimada, T., Kuwabara, K., Yamakoshi, S., & Shichi, T. (2002). A case of infanticide in the spot-billed duck in circumstances of high breeding density. *Journal of Ethology*, *20*, 87–88.

Shostak, M. (1981). *Nisa: The life and words of a !Kung woman.* Cambridge, MA: Harvard University Press.

Sicotte, P. (2002). The function of male aggressive displays towards females in mountain gorillas. *Primates*, *43*(4), 277–289.

Silk, J. (1994). Social relationships of male bonnet macaques: Male bonding in a matrilineal society. *Behaviour*, *130*(3–4), 271–291.

Silvered leaf-monkey (*Trachypithecus cristatus*). (2008). Retrieved from http://www.theprimata.com/trachypithecus_cristatus.html

Smith, D. (2007). *The most dangerous animal: Human nature and the origins of war.* New York: St Martin's Press.

Smuts, B. (1985). *Sex and friendship in baboons.* New York: Aldine Publishing.

Sommer, V. (2000). The holy wars about infanticide. Which side are you on? And why? In C. van Schaik & C. Janson (Eds.), *Infanticide by males and its implications* (pp. 9–26). Cambridge: Cambridge University Press.

Sparks, J. (1967). Allogrooming in primates: A review. In D. Morris & D. Bruce (Eds.), *Primate ethology* (pp. 148–175). London: Weidenfeld and Nicolson.

Stanford, C. (1998). The social behavior of chimpanzees and bonobos: Empirical evidence and shifting assumptions. *Current Anthropology*, *39*(4), 399–407.

Stanford, C. (1999). *The hunting apes: Meat eating and the origins of human behavior.* Princeton, NJ: Princeton University Press.

Stauffer, R., Walker, A., Ryder, O., Lyons-Weiler, M., & Hedges, S. (2001). Human and ape molecular clocks and constraints on paleontological hypotheses. *Journal of Heredity*, *92*, 469–474.

Stein, D. (1984a). Ontogeny of infant-adult male relationships during the first year of life for yellow baboons (*Papio cynocephalus*). In D. Taub (Ed.), *Primate paternalism* (pp. 213–243). New York: Van Nostrand Reinhold.

Stein, D. (1984b). *The sociobiology of infant and adult male baboons.* Norwood, NJ: Ablex Publishing.

Steudel-Numbers, K. (2006). Energetics in *Homo erectus* and other early hominins: The consequences of increased lower-limb length. *Journal of Human Evolution*, *51*, 445–453.

Stewart, K. (2003). *Gorillas: Natural history and conservation.* Stillwater, MN: Voyageur Press.

Strier, K. (1992). *Faces in the forest: The endangered muriqui monkeys of Brazil.* New York: Oxford University Press.

Strier, K. B., Dib, L. T., & Figueira, J. E. C. (2002). Social dynamics of male muriquis (*Brachyteles arachnoides hypoxanthus*). *Behaviour, 139*(2–3), 315–342.

Strum, S. (1987). *Almost human: A journey into the world of baboons.* New York: Norton.

Sugiyama, Y. (1967). Social organization of hanuman langurs. In S. Altmann (Ed.), *Social communication among primates* (pp. 221–236). Chicago: University of Chicago Press.

Sugiyama, Y. (1971). Characteristics of the social life of bonnet macaques (*Macaca radiata*). *Primates, 12*(3–4), 247–266.

Sumner, P., & Mollon, J. (2003). Colors of primate pelage and skin: Objective assessment of conspicuousness. *American Journal of Primatology, 59*(2), 67–91.

Sussman, R., & Cloninger, C. (Eds.). (2011). *Origins of altruism and cooperation.* New York: Springer.

Sussman, R., & Garber, P. (2011). Cooperation, collective action, and competition in primate social interactions. In C. Campbell, A. Fuentes, et al. (Eds.), *Primates in perspective* (2nd ed., pp. 587–599). New York: Oxford University Press.

Sussman, R. W., Garber, P. A., & Cheverud, J. M. (2005). Importance of cooperation and affiliation in the evolution of primate sociality. *American Journal of Physical Anthropology, 128*, 84–97.

Sussman, R., & Marshack, J. (2010). *Are humans inherently killers? A critique by the authors followed by a response by Richard Wrangham* (Global Nonkilling Working Papers, ISSN 2077–1428). Honolulu: Center for Global Nonkilling.

Suwa, G., Kono, R. T., Simpson, S. W., Asfaw, B., Lovejoy, C. O., & White, T. D. (2009). Paleobiological implications of the *Ardipithecus ramidus* dentition. *Science, 326*, 69.

Takahata, Y. (1982). Social relations between adult males and females of Japanese monkeys in the Arashiyama B troop. *Primates*, *23*(1), 1–23.

Tanner, N. (1981). *On becoming human.* Cambridge: Cambridge University Press.

Tardif, S., Santos, C., Baker, A., Van Elsacker, L., Feistner, A., Kleiman, D., Ruiz-Miranda, C., et al. (2002). Infant care in lion tamarins. In D. Kleiman & A. Rylands (Eds.), *Lion tamarins: Biology and conservation* (pp. 213–232). Washington, DC: Smithsonian Institution Press.

Tattersall, I., & Mearns, J. (2000). Once we were not alone. *Scientific American*, *282*, 56–62.

Taub, D. (1984). Male caretaking behavior among wild barbary macaques (*Macaca sylvanus*). In D. Taub (Ed.), *Primate paternalism* (pp. 20–55). New York: Van Nostrand Reinhold.

Taylor, H., Teas, J., Richie, T., Southwick, C., & Shrestha, R. (1978). Social interactions between adult male and infant rhesus monkeys in Nepal. *Primates*, *19*(2), 343–351.

Teichroeb, J., & Sicotte, P. (2010). The function of male agonistic displays in ursine colobus monkeys (*Colobus vellerosus*): Male competition, female mate choice or sexual coercion? *Ethology*, *116*, 366–380.

Thornhill, R., & Palmer, C. (2000). *A natural history of rape: Biological bases of sexual coercion.* Cambridge, MA: MIT Press.

Thorpe, S., Holder, R., & Crompton, R. (2007). Origin of human bipedalism as an adaptation for locomotion on flexible branches. *Science*, *316*, 1328.

Thorpe, I. J. N. (2005). *The ancient origins of warfare and violence.* Paper presented at the Conference on warfare, violence and slavery in prehistory, Sheffield University.

Tierney, P. (2000). *Darkness in Eldorado: How scientists and journalists devastated the Amazon.* New York: Norton.

Tiger, L. (1969). *Men in groups.* New York: Random House.

Tiger, L., & Fox, R. (1971). *The imperial animal.* New York: Holt, Rinehart and Winston.

Toft, S. (1986). Domestic violence in Papua New Guinea (Law Reform Commission Occasional Paper 19). Port Moresby, New Guinea: Papua New Guinea Law Reform Commission.

Toups, M. A., Kitchen, A., Light, J. E., & Reed, D. L. (2011) Origin of clothing lice indicates early clothing use by anatomically modern humans in Africa. *Molecular Biology and Evolution, 28*(1), 29–32.

Treves, A. (1997). Primate natal coats: A preliminary analysis of distribution and function. *American Journal of Physical Anthropology*, *104*(1), 47–70.

Udell, M., Dorey, N., & Wynne, C. D. L. (2011). Can your dog read your mind? Understanding the causes of canine perspective taking. *Learning and behavior*, *39*(4), 289–302.

United Nations Children's Fund & UNAIDS Secretariat (2006). *Africa's Orphaned and Vulnerable Generations: Children Affected by AIDS.* New York: UNICEF and UNAIDS.

Valeggia, C. R., Mendoza, S. P., Fernandez-Duque, E., Mason, W. A., & Lasley, B. (1999). Reproductive biology of female titi monkeys (*Callicebus moloch*) in captivity. *American Journal of Primatology*, *47*, 183–195.

Van Hooff, J., & van Schaik, C. (1994). Male bonds: Affiliative relationships among nonhuman primate males. *Behaviour*, *130*(3–4), 309–337.

Van Lawick-Goodall, J. (1971). *In the shadow of man*. Boston: Houghton Mifflin.

Van Noordwijk, M., & van Schaik, C. (2000). Reproductive patterns in eutherian mammals: Adaptations against infanticide? In C. van Schaik & C. Janson (Eds.), *Infanticide by males and its implications* (pp. 322–360). Cambridge: Cambridge University Press.

van Schaik, C. (2000). Social counterstrategies against infanticide by males in primates and other mammals. In P. Kappeler (Ed.), *Primate males: Causes and consequences of variation in group composition* (pp. 34–52). Cambridge: Cambridge University Press.

van Schaik, C., & Dunbar, R. (1990). The evolution of monogamy in large primates: A new hypothesis and some crucial tests. *Behaviour, 115*, 30–62.

van Schaik, C., & Kappeler, P. (1997). Infanticide risk and the evolution of male-female association in primates. *Proceedings of the Royal Society B: Biological Sciences, 264*, 1687.

Veiga, J. (1990). Infanticide by male and female house sparrows. *Animal Behaviour, 39, 496–502.*

Veiga, J. (2004). Replacement female house sparrows regularly commit infanticide: gaining time or signaling status? *Behavioral Ecology, 15*(2), 219–222.

Vick, L. (2008). Immaturity in spider monkeys: A risky business. In C. J. Campbell (Ed.), *Spider monkeys: Behavior, ecology and evolution of the genus* Ateles (pp. 288–328). Cambridge: Cambridge University Press.

Videan, E., & McGrew, W. (2001). Are bonobos (*Pan paniscus*) really more bipedal than chimpanzees (*Pan troglodytes*)? *American Journal of Primatology, 54*(4), 233–239.

Vines, G. (1999). Queer creatures. *New Scientist, 2198*, 33–35.

Wade, N. (2006). *Before the dawn: Recovering the lost history of our ancestors*. New York: Penguin Press.

Walker, P. L. (2001). A bioarchaeological perspective on the history of violence. *Annual Review of Anthropology, 30*(1), 573–596.

Walter, C. (2006). *Thumbs, toes, and tears, and other traits that make us human.* New York: Walter and Co.

Ward, C. V., Kimbel, W. H., & Johanson, D. C. (2011). Complete fourth metatarsal and arches in the foot of *Australopithecus afarensis*. *Science, 331*, 750–753.

Washburn, S., & DeVore, I. (1961). Social behavior of baboons and early man. In S. Washburn (Ed.), *Social life of early man* (pp. 91–105). Chicago: Aldine Publishing.

Watson, E. E., Easteal, S., & Penny, D. (2001). *Homo* genus: A review of the classification of humans and the great apes. In P. Tobias, M. Raath, et al. (Eds.), *Humanity from African naissance to coming millennia: Colloquia in human biology and palaeoanthropology* (pp. 307–318). Firenze, Italy: Firenze University Press.

Weber, B., & Vedder, A. (2001). *In the kingdom of gorillas: Fragile species in a dangerous land.* New York: Simon and Schuster.

Weiss, K., & Buchanan, A. (2009). *The mermaid's tale: Four billion years of cooperation in the making of living things.* Cambridge, MA: Harvard University Press.

White, T. D., Asfaw, B., Beyene, Y., Haile-Selassie, Y., Lovejoy, C. O., Suwa, G., & WoldeGabriel, G. (2009). *Ardipithecus ramidus* and the paleobiology of early hominids. *Science*, *326*, 64.

Whiten, A., Goodall, J., McGrew, W., Nishida, T., Reynolds, V., Sugiyama, Y., Tutin, C. (2001). Charting cultural variation in chimpanzees. *Behaviour*, *138*, 1481–1516.

Whitten, P. (1986). Infants and adult males. In B. Smuts et al. (Eds.), *Primate societies* (pp. 343–357). Chicago: University of Chicago Press.

Wilson, E. (1975). *Sociobiology: The new synthesis*. Cambridge: Harvard University Press.

Winter, H., Langbein, L., Krawczak, M., Cooper, D. N., Jave-Suarez, L. F., Rogers, M. A., Praetzel, S., et al. (2001). Human Type 1 hair keratin pseudogene *ψhHaA* has functional orthologs in the chimmpanzee and gorilla: evidence for recent inactivation of the human gene after the Pan-Homo divergence. *Human Genetics, 108*, 37–42.

Wittig, R. (2010). The function and cognitive underpinnings of post-conflict affiliation in wild chimpanzees. In E. Lonsdorf, S. Ross, et al. (Eds.), *The mind of the chimpanzee: Ecological and experimental perspectives* (pp. 208–219). Chicago: University of Chicago Press.

Wolf, K. (1984). *Reproductive competition among co-resident male silvered leaf monkeys (Presbytis cristata)* (Unpublished doctoral dissertation). Yale University, New Haven, CT.

Wolf, K., & Fleagle, J. (1977). Adult male replacement in a group of silvered leaf-monkeys (*Presbytis cristata*) at Kuala Selangor, Malaysia. *Primates, 18*, 949–955.

Woods, V. (2010). *Bonobo handshake: A memoir of love and adventure in the Congo.* New York: Gotham Books.

Wrangham, R. (1974). Artificial feeding of chimpanzees and baboons in their natural habitat. *Animal Behaviour, 22*, 83–93.

Wrangham, R. (2009). *Catching fire: How cooking made us human.* New York: Basic Books.

Wrangham, R., & Peterson, D. (1996). *Demonic males: Apes and the origins of human violence.* Boston: Houghton Mifflin.

Yamagiwa, J. (1987). Intra- and inter-group interactions of an all-male group of Virunga mountain gorillas (*Gorilla gorilla beringei*). *Primates, 28*(1), 1–30.

Yerkes, R. (1925). *Almost human.* New York: Century Co.

York, G. (2011, September 3). Nomination for South Africa's chief justice provokes outrage from lawyers, women's groups. *Globe and Mail*, p. A14.

Zhou, J., Wei, F.-w., Li, M., Lok, C. B. P., & Wang, D. (2008). Reproductive characters and mating behaviour of wild *Nomascus hainanus. International Journal of Primatology, 29*, 1037–1046.

Zihlman, A. L., Cronin, J. E., Cramer, D. L., & Sarich, V. M. (1978). Pygmy chimpanzee as a possible prototype for the common ancestor of humans, chimpanzees and gorillas. *Nature, 275*, 744–746.

Zuckerman, S. (1932). *The social life of monkeys and apes.* London: Kegan Paul, Trench, Trubner.

Index

About the Authors

Anne Innis Dagg is a faculty member in the Independent Studies program at the University of Waterloo. She holds a PhD from the University of Waterloo and a BA and MA from the University of Toronto. Dr. Dagg has published over 60 refereed papers related to animal behavior and bias in academia. Her recent books include *Love of Shopping is Not a Gene: Problems with Darwinian Psychology*, *Pursuing Giraffe*, *The Social Behavior of Older Animals*, and *Animal Friendships*.

Lee E. Harding is an environmental consultant with a PhD in Wildlife Toxicology from Gifu University (Japan) and a BSc in Wildlife Management from Humboldt State University (California). Dr. Harding is a Registered Professional Biologist (R.P.Bio.) in British Columbia, Canada. As an environmental consultant from 1972 to 1976, he studied wildlife ecology in forest and tundra ecosystems throughout northern and western Canada. In 1976 he joined Environment Canada to manage pollution assessment programs, first in Alberta and Northwest Territories and then in British Columbia. From 1994 to 1997 he managed the Wildlife Toxics group at the Canadian Wildlife Service in Delta, BC. In 1998 he returned to environmental consulting, studying wildlife in many parts of the world. In his spare time, he writes on toxicology and conservation biology of endangered species with a recent focus on primates in Southeast Asia.